Math *Is* Hard

But Ace it Anyway

Math *Is* Hard

But Ace it Anyway

Rebecca DeCamillis, B.Sc.

Order this book online at www.trafford.com
or email orders@trafford.com

Most Trafford titles are also available at major online book retailers.

Printed in Victoria, BC, Canada.

ISBN: 978-1-4269-1579-6 (Soft)

*Our mission is to efficiently provide the world's finest, most comprehensive
book publishing service, enabling every author to experience success.
To find out how to publish your book, your way, and have it available
worldwide, visit us online at www.trafford.com*

Trafford rev. 03/25/2010

 www.trafford.com

North America & international
toll-free: 1 888 232 4444 (USA & Canada)
phone: 250 383 6864 ✦ fax: 812 355 4082

For Colin, with all my love.

And for my students, past and present,
who have taught me so much of what I know.

Contents

Introduction

Hi, everyone! I am so glad you chose to read this book because it is going to be such an adventure for you. It may even radically change the way you think and feel about math. This is a totally different kind of math book because instead of teaching you how to do math, it will to teach you how to *learn* math and how to *like* math. You are going to love it!

I will start by sharing with you my personal journey from almost giving up hope on math in high school to teaching it today. I want to share this because it is important for you to know that success in math is not out of your reach. My goal in writing this book is to convince you that if you change your habits and thinking, maybe even a little, you can be awesome at math. I want to be an inspiration to you, because if I got through math—trust me—so can you.

So here's my story. When I was in elementary school, though I was never a genius or even especially gifted in math, I usually did well in the subject and even enjoyed it a little. But as the years went on, I learned to like the subject less and less. I know now that I grew to dislike it so much because I didn't understand it. And I had no idea how to help myself understand it; I did not know how to study. What a terrible feeling it is to know that even though you are trying, you still cannot do well. I now understand that one of the challenges with math is that effort counts for very little when it comes to tests. You either know the material or you don't—and I usually didn't.

In high school, I struggled tremendously with the subject. By the time I reached Grade 10, I dreaded going to math class and doing the homework. I felt stuck and helpless, and was constantly frustrated. In fact, I developed such an anxiety about math that I had to write my tests in another classroom just to be able to get over my nervousness and concentrate. Looking back, I see that even that helped me little, as I passed Grade 10 math with a C-.

In Math 11, I had an even harder time. I did my best in class and with homework, but I still only passed by a few percent. As the end of the year approached, the teacher told me he thought I was dyslexic and that I would be crazy to take Math 12. If I remember correctly, he recommended I take an adapted Math 11 course the following year.

I ignored his advice and took Math 12. To my disbelief, the most incredible thing happened: I ended up with a C+ in the course. For someone who had struggled the two previous years just to barely pass, this grade was outstanding. So what had changed?

Well, for the first time in high school I had a math teacher I liked. He was energetic and funny and caring and supportive, and I found myself wanting to try harder. He taught in a way that I could actually understand, and for the first time in high school, I started to believe in myself (in math anyway!).

I was also motivated to do better because I knew that Math 12 was important if I wanted to go to college, which I did. But one of the best things that happened to change the way I viewed math that year was finding a friend who also struggled in math.

My attitude completely changed and I worked even harder than usual. I went to school early most mornings to get extra help from my teacher, and spent at least an hour most nights studying and doing homework. At least a couple of nights a week, my friend and I had joint study sessions, helping and supporting and coaching each other, laughing together, and even figuring out some math. In class, the two of us sat right at the front. I had never sat in front in a math class; in fact, I had always made a conscious effort to sit as close to the back of the room as possible.

I was starting to feel happy again. I remember how great it felt to actually understand what was going on in class and to have the support I had needed for so long. This great turning point in my math experience was due to the combination of having a study partner, an amazing teacher, and the will to succeed.

Near the end of Grade 12, I applied at Simon Fraser University but was not accepted; my grades were not high enough for the Science program. I was disappointed but did not give up. Instead, I applied at Douglas College (a local community college). They accepted me, and I registered in the General Sciences program. At the time, I wanted to become a chiropractor. I loved the idea of working with people and being able to help them with just my mind and my hands, and this seemed like an exciting and lucrative occupation.

For the General Sciences program I was required to take two semesters of calculus. I took a precalculus course my first semester to prepare me for the real thing, and I did well—it was basically the same as Math 12. I took the first actual calculus course the following semester. Calculus turned out to be hard. And since I was taking it along with four other really challenging courses that semester, I ended up withdrawing from the calculus course just a few weeks before the end of the semester. There was no way I was going to pass.

Clearly calculus was not my area of expertise, but I knew in my heart that I had to get through it. I was not going to give up. So rather than just taking it again in the fall (and risking failing it), I hired a math tutor over the summer, who took me through the entire calculus course.

Investing my summer toward learning that course properly was one of the smartest things I ever did during my post-secondary education. It was extremely helpful because I was able to learn the material at my own pace with no tests to worry about. And when I retook the course in the fall, I got a B.

Things were finally looking up. Not only did I get a good grade, but I also actually enjoyed the class a little. I gained confidence because I was doing amazingly on my tests (compared to what I'd been used to). I also began to like the coursework because I finally understood it. And guess what? I learned that once you understand math, it's actually kind of interesting.

The following semester I took the second calculus course and did fairly well. So I reapplied to Simon Fraser University for the summer semester, and was accepted! Being admitted to SFU was an emotional and joyful experience for me, because I had always believed the university was beyond my reach. In fact, upon my acceptance, I had remembered driving by SFU a few years earlier with some friends and feeling so disappointed and upset, because I was sure I would never have the opportunity to go there.

That summer, I attended SFU, but with a brand new career goal. I had decided that becoming a chiropractor would require an extremely long time in school with no guarantee that I would even be good at it or like it. Unwilling to take that kind of risk, I had a long talk with a friend about my situation. I told her that I knew I still wanted to help people but was unsure how. After a lot of coffee and much thought, I decided on teaching high school. I knew I would enjoy teaching older students more, as I would be better able to relate to them. Then I thought about what subject I might like to teach. And through a process of elimination, I decided on math.

My first semester at Simon Fraser was tough. I took too many math courses and wasn't working nearly hard enough, and as a result I almost failed a class. I was placed on academic probation for the following semester, which meant that if my grades did not improve, I would be forced to withdraw from the school. This really scared me. If I had to leave SFU, all of my hard work at getting in would have been for nothing.

Luckily, I learned from my mistakes. I lightened my course load the following semester and kept it that way, never taking more than two math courses at once. This allowed me to spend more time on my math courses and also to have some variety. I took several philosophy and English courses along with the math, and really enjoyed them. My grades improved that semester, and the university let me stay.

After that, I got serious. I started to really focus, and I found an awesome tutor who later became a great friend. Things started to go more smoothly. I was starting to understand the higher levels of math more and was finding it interesting. It was still hard, but the combination of my effort, having friends there to study with, and having a good tutor seemed to be enough to give me some success. I was adapting to the lifestyle and was beginning to create a life for myself as a student.

About a year later, my tutor told me about a learning center he worked for that was looking to hire a math tutor. I was interviewed for the position and hired the same day. I could not believe that I, of all people, got a job teaching math! Kind of funny, don't you think? Once

I learned the job, though, I loved it. The students seemed to like me, and they were learning and improving quickly. I got along with them well, especially the teenagers, which confirmed for me that my career choice was the right one. And I found that understanding where they were having difficulty came naturally to me; struggling so much in math had made teaching it easy!

The more I worked at the learning center, the more I enjoyed it and the more convinced I became that teaching math was my calling in life. I began trying harder than ever at school. For the next couple of years, I devoted most of my life to earning my degree: a bachelor of science with a major in mathematics. My tutor, along with friends, professors, teacher's assistants, and anyone else I could find, continued to help me until I graduated. I spent hours upon hours trying to figure out math, and sure enough, I eventually earned my degree. I was overjoyed; this was the biggest accomplishment of my life.

Shortly after graduation, I was accepted into the teaching program at SFU. This was a one-year program that, for the most part, I enjoyed. The six-week student-teacher practicum was unbelievably challenging. This is the part of the teaching program in which the teacher-in-training actually takes on the entire workload of a regular teacher. In general, however, I found the teaching program much easier than the math degree. For the first time in years, I didn't have to worry about failing tests or not understanding. It was so refreshing to be doing something different, something that came more naturally to me, and before I knew it, it was all over. I had made it—I was a math teacher!

Now, ten years later, I teach and tutor high school math, and I absolutely love it. And as mentioned before, I am writing this book to share some of what I learned along the way, in the hope that it can help make your journey through high-school math easier and less stressful than mine was. I want you to leave high school as a confident young adult, knowing that you are smart and capable and that you understand math.

Do you know that probably 80% or more, of the adults I meet tell me they didn't like or could never do math in high school? There is no way that all of these people are incapable of learning the subject, so there must another reason. I believe the reason is that a majority of high school students are not taught one of the most important things: how to *learn* math and be successful at it!

I don't want you to be one of these people. I want you to be one of the shining few who graduate from high school with memories of liking and doing well in math. Make a serious effort to apply the strategies and ideas presented in this book, and I promise that you will not only become confident and strong in math, but will also start to enjoy it!

So let's see if I can help. Enjoy the book.

Chapter 1: The Top 10 Self-Destructive Habits of the Typical Math Student
(In No Particular Order)

A self-destructive habit is any habit a person develops that causes them harm either immediately or later on in life. Most people develop these habits because the behavior is comforting or fun or "the easy way" at the time. Consider smoking, for example, or overeating or overspending or not doing your homework. All of these indulgences provide what is referred to as "immediate gratification," which basically means they help you to feel good right now. The problem with such behaviors is that, in the end, they will inevitably create problems.

In this chapter, we will look at the most common self-destructive habits that I have observed among math students, as well as why these habits are so harmful. As you read, consider whether you display any of these behaviors.

My goal in helping you become aware of such habits is not to make you feel bad about yourself, but to help you fix anything that may be causing you problems. After all, the only way to get better at something is to first be aware that it is a problem and then recognize the specific area that needs improvement. Think about an athlete who wants more than anything to win a marathon. Suppose she has run fifty marathons and has never finished in the top five. She may just need to learn a new skill or training method; for example, maybe she should do more of a certain leg exercise or add something different to her diet. But whatever the problem, she will never be able to correct it if she does not know what it is. The same is true for math. If you do not identify what you might be doing to cause yourself stress or to keep yourself from getting the results you want, then the problem will never go away.

In my ten years of teaching, I have found that the following ten habits cause students the most trouble. In fact, most students who do poorly in math practice at least *half* of these habits.

So as you read, put a checkmark beside any of these that you think may apply to you. Then look back in a few months and see if you have made some progress.

Habit #1: Not using your notes and textbook

Did you know that the whole purpose of taking notes is to create a study tool for yourself, to help you do homework and study for tests? They are a tool for you. They are there to *help you*.

So many students take notes without thinking about what they are writing, and then never look at them again. That seems like an awful waste of time. If this sounds like you, next time you take notes, ask yourself, "What is the point of taking notes and never using them?" Well-organized notes are an excellent study tool. If you develop the habits of writing your notes in a way that you can understand them, and reading them, you will see that they can help you understand your homework and prepare for tests and quizzes. So begin taking notes in a way that you *can* use them later; write neatly and keep them organized, and then read them!

Just as valuable is reading your textbook. Before you start your homework, read or reread the corresponding section in your textbook to gain a better grasp of the ideas. Then take a few minutes to thoroughly look over the examples; they can be extremely helpful. As you do both of these things, pay close attention to anything in boldface, or anything with a border or a box around it, as important ideas are usually presented in this way.

You may have already found that math textbooks can be quite difficult to read. They are dense, which means that there is usually a lot of information in each paragraph. So take your time, read slowly, and be patient with yourself. Look up any words you do not understand, either in the glossary or in a dictionary or on the Internet. You may also be able to use your notes to help you understand certain terms or concepts in the textbook. If there are still some parts of the reading that you continue to struggle with, ask a teacher or your tutor for help.

Habit #2: Not listening to or getting involved in the lesson
I understand that listening to a math lesson is not always as fun as talking to the friend who sits in front of you in class. Neither is it as fun as listening to music or playing a game on your calculator or daydreaming about the weekend. But here is the bottom line: Not everything in life will be really fun all the time, and sometimes if you try hard at something that is not especially fun, it will make life easier for you later. In math, this means that if you get involved in the lesson, it is more likely that you will understand the concepts, that the homework will become easier, that your grades will be higher, and that you will feel better about yourself.

Asking questions is a crucial part of participating in a lesson, because it is the only way you can get clarification on something you do not understand. Suppose the teacher introduces an idea or writes out a step you don't understand. If you do not ask about it, chances are you will only get more and more confused as the lesson goes on. But if you ask, you will be more likely to understand the rest of the lesson. If you are still confused even after you ask your question, be honest and ask for more clarification. There are probably other students who are as confused as you are!

Remember that being able to go to classes and have a teacher explain things to you is a *gift;* the whole point of lessons is to help you learn. Participating in class will make your homework easier and you happier because it will help you understand the math. On the other hand, if you choose not to pay attention or get involved, you will probably go home and feel frustrated because you have no idea how to do your homework. So the next time you want to talk to that friend in front of you during class ask yourself, "Is it worth it?"

Habit #3: Showing up late for class or leaving during the lesson
Of course life happens sometimes and coming late to school is unavoidable. The problem is that if you come late to class, you risk missing the main idea of the lesson, which means you may spend the rest of class feeling confused and trying to catch up. If you have trouble

with math, showing up late on a regular basis will make things much harder for you.

Leaving before the teacher has finished the lesson—even if just for a bathroom break—can cause just as many problems for you. Teachers usually present and discuss challenging examples during lessons that help students understand the homework. If you leave during this time, you risk having trouble with the more challenging homework questions.

So try your best to arrive on time. Then sit and listen and ask questions. And if you absolutely need to leave the class, try to wait until after the lesson.

Habit #4: Relying too much on your calculator

When students use their calculators too often, they weaken their ability to do simple math. Several of the adolescents I tutor (including Grade 12 students) have almost entirely forgotten their multiplication tables, as well as how to do simple calculations such as 0.5×2. It only takes a minute of thought to recall that 0.5 is 1/2, and 2 halves are one, so $0.5 \times 2 = 1$. These students do not perform such simple calculations mentally because using the calculator is easier.

The problem: if you use your calculator to do all of the work for you, you will never get better at doing it on your own. Your ability to work with numbers cannot improve without practice. Relying so heavily on your calculator is like spending all of your time in a wheelchair when your legs are perfectly fine, just because you don't feel like walking! If you did this for long enough, eventually your legs would get so weak that walking will become difficult.

Another problem is that calculators are finicky. You have to enter everything perfectly, especially calculations that require brackets, or the answer will be wrong. Suppose you had to evaluate the following math problem:

$$(-1)^2 + \frac{7 \times 0 + 5}{5} - \frac{4}{2}$$

In a typical scientific calculator, this would have to be entered as follows:

$(-1) x^2 + ((7 \times 0 + 5) \div 5) - (4 \div 2) =$

There are a number of ways to enter this incorrectly; most students make mistakes with the brackets. But even if you do not make a mistake, entering something like this into the calculator is time-consuming. On the other hand, with practice, you could determine the answer much faster mentally, by evaluating it one term at a time like this:

$$(-1)^2 + \underset{\underset{\textbf{2.}}{\downarrow}}{\frac{7 \times 0 + 5}{5}} - \underset{\underset{\textbf{3.}}{\downarrow}}{\frac{4}{2}}$$

$\underset{\underset{\textbf{1.}}{\downarrow}}{}$

1. $(-1)^2 = 1$

2. $\dfrac{7 \times 0 + 5}{5} = \dfrac{5}{5} = 1$

3. $\dfrac{4}{2} = 2$

4. And putting it all together we get $1 + 1 - 2 = 0$

See? You can figure this out without a calculator, and the math involved is not hard—all you have to do is remember the correct order of operations. So to summarize: calculators can weaken your math skills and are often more of a pain than a help.

The purpose of calculators is to save you time so you are not spending hours multiplying complicated decimals or performing grueling long division—*but anything that can be done in your head should be done in your head.* The more you do math mentally, the stronger and quicker you will become at it.

A final note about calculators: Many of the students who over-use their calculators also over-trust their calculators. What I mean is, the student will automatically trust the answer showing on the calculator even if it is unreasonable. For example, suppose a student has to calculate 50 ÷ 13. He enters it into the calculator and gets 38.46 for an answer. So he writes this answer down and moves onto the next question. What went wrong here? Well, 38.46 is the answer to 500 ÷ 13, so the student accidentally entered 500 into the calculator instead of 50. This is a common error and easy to correct, but this student could not correct it because he failed to notice that something about the answer was strange: how can 13 possibly go into 50, 38 times?

If over-trusting your calculator is one of your bad habits, all you have to do to correct it is, each time you get an answer on your calculator ask yourself whether the answer seems reasonable. This will work much better for you. Learn to use your calculator and think at the same time!

Habit #5: Waiting until the last minute to start homework

In high school, this was my worst bad habit. I did not like math at all, so I would procrastinate until I absolutely had to do the homework. If it was due on Monday morning, I would start it late Sunday evening. This habit caused many problems for me, and created a lot of extra stress and anxiety in my life.

I always dreaded Sundays because I knew in the back of my mind that at some point I would have to face my homework. Sunday is supposed to be a day for rest, and I turned it into a day of stress because I never learned to push myself to get the work done sooner. Imagine how much better off I would have been if I had completed my homework on Friday after school instead.

Doing homework at the last minute also meant that by the time I actually started the homework, I had usually forgotten the lesson—and, of course I never read my notes—so I would feel overwhelmed, not even knowing where to start. Inevitably, by 9:00 at night, I would almost always be stressed out and frustrated, with no time to get help.

The best idea is to finish your homework before you even leave school. Try to use class time to do it, if your teacher allows any. If you are able to do only a portion of it in class, spend the majority of your time working on the harder questions instead of doing them in order.

This way you can get help from the teacher if you need it, rather than doing the easier ones first and running out of time, only to get stuck on the more challenging ones when you get home. This strategy may work better for you as your only homework will be the easier questions (and to review the more challenging ones).

Habit #6: Not studying enough or properly for tests and quizzes
Many students do their homework but then do not study enough for tests, if at all. Yet, in most math classes, tests and quizzes are by far the most heavily weighted part of the course, making up at least 80% of the final grade. Homework, on the other hand, only counts for about 15% to 20% of the final grade.

What this means is that if you don't invest time in studying, you will likely get a poor grade on the test, which will probably make you dislike math even more. If, however, you work especially hard for the few days leading up to the test, you will likely do well—and then you will feel happy and want to keep doing well! Preparing well for tests and quizzes is one of the most important things you can do for yourself if you want to be successful in math.

Note: If you do study but are still not getting the grades you want, chances are you are not studying the right way. And yes, there are right and wrong ways to study math, but I will go into more detail about that later.

Habit #7: Memorizing or copying steps or examples instead of trying to understand the math
This is one of the most self-destructive habits to develop, mainly because most of the enjoyment of math comes from understanding it, but also because often students memorize and copy steps as a way to avoid learning the math.

Concerning the first point, there is great satisfaction in knowing that you figured out a math problem, but where is the joy in following a prescribed set of steps that you do not understand, only to arrive at an answer that has no meaning to you? In George Polya's *How To Solve It*, he writes,

> It is foolish to answer a question that you do not understand. It is sad to work for an end that you do not desire. Such foolish and

> sad things often happen, in and out of school, but the teacher should try to prevent them from happening in his class. The student should understand the problem. But he should not only understand it, he should also desire its solution. (6)

When solving a math problem, your goal should be to understand it, not just to get the right answer. Think about approaching it the way you would a tricky riddle: with curiosity and interest. If you can read the problem enough so you know what is asking for, and then develop a genuine desire to figure it out, you will be more aware of what the steps mean, and you will be less likely to end up with answers that do not make sense. And once you arrive at the solution, you will be much happier because it will mean something to you. Unfortunately, this method of doing math is harder, and it requires more thought, more time, more effort, and potentially more frustration. It is much easier to copy or memorize steps. The problem is if you do not challenge yourself and try to learn the math, you will not grow and get stronger in math, which means as time goes on, math will become increasingly difficult for you.

Many students do survive the first few years of high school math by simply memorizing steps or solutions; unfortunately, usually they are not learning the concepts behind the steps, or even the reasons for the steps.

Memorizers may even go so far as to copy the exact steps from a previous question or from an example in the notes, even if the steps do not apply to the problem they are working on. This is the danger of not understanding the math: if you copy the steps of a previous question and the previous question is even slightly different than the one you are working on, you will end up with the wrong answer. Students often do this when they encounter a word problem; they will look back at the notes to find a similar problem and then copy down the steps from that problem without understanding what they wrote. Sometimes the steps they write down do not even make sense; all they know is that the questions looked the same, so the steps should work the same. I know, because I did this very thing! One of my students tried it the other day. The problem she had in her notes was the following:

Two numbers have a difference of 10 and their product is a minimum. Find the numbers.

There are two main equations involved in the solution to this problem. If x and y are the numbers we are supposed to find, then the first equation would be $x - y = 10$. My student had to solve the following problem:

Two numbers have a sum of 10 and their product is a maximum. Find the numbers.

Since this question looks similar, she began with the equation $x - y = 10$. But the problem she was working on states that the sum of the numbers is 10, not the difference, so the equation should have been $x + y = 10$. This small change means the difference between the right answer and the wrong answer.

If she had read the problem carefully first and tried to understand it, she probably would have noticed that it doesn't make sense to use the same equation in both problems.

So why do students do this? They do it because they don't know what else to do (or they don't feel like thinking, which is also common). This is what they have always done in the past and it has worked. They have not developed *mathematical independence*, which means they have not learned how to think about math and come up with creative solutions. What's sad is that this way of doing math is detrimental to the student's attitude and feelings about the subject. Students who rely on memorizing or copying steps feel uneasy with the process, even after they finish their homework, because they know deep inside that they truly don't get it. Plus, these students are missing out on the awesome satisfaction that comes from understanding math.

The longer students go on like this, the worse their situations become. As math progresses, the questions get harder and less repetitive, and there is often no longer a recipe to follow. At this point—usually by the second or third year in high school—students need to be able think on their own, ask questions, and use problem-solving strategies. It is at this stage, for the memorizer, that math can suddenly seem almost impossible to do. The memorization techniques that have worked for the student so far suddenly stop working—and the student often will not know how to improve the situation, because she will not have learned how to *think* about math. Also, memorizers will likely lack the knowledge base that should have been developed over the previous years, because they will have never really learned those concepts to begin with. It is no coincidence that most adults remember this period in high school as being the time they really began to dislike math!

If this experience sounds like yours, then make a major change right away! Start by shifting your focus during homework time from just finishing the work to really understanding it, and whenever you have trouble understanding, ask for help. Also, though sometimes math seems too hard to understand, and memorizing and following steps may seem the only way, it's actually worth it to *not* do the homework until you get help. Remember, you may get good marks by completing assignments, but homework only accounts for a small percentage of your final grade. Tests are what really count—and to be successful on a test, you must understand the math.

In summary, remember two things:

- To do well in and enjoy math, you have to develop your own mathematical independence, which means thinking hard and working to understand the concepts. You establish this independence by understanding, analyzing, and solving problems; using your mind and your skills, not by memorizing a set of steps. In other words, there are no good shortcuts.

- Math is supposed to make sense. To grow in math and to enjoy it, you must know why you are doing what you're doing for each and every question. This is the only way to gain confidence and ability.

The bottom line: learn the math so you can get better at it, do it on your own, enjoy it, and ultimately feel good about yourself.

A few more words about memorization

There are situations when memorization is okay. For example, you do not always need to know exactly why every formula or algorithm works (although it is helpful), but you do need to understand what they mean and when to use them.

Suppose you read a word problem and the solution involves multiplying fractions. If you have the multiplication method memorized but do not know exactly why it works, that is all right. The important thing is to recognize what the problem is about or what math concept it involves—in this case, you would need to know that the problem is about multiplying fractions.

Here is a more specific example of what I mean. Say you are asked to determine how much air will fill a beach ball. You may not know (and you are not expected to know) *why* the formula for the volume of a sphere is $V = \frac{4}{3}\pi r^3$, but you do need to be able to recognize that this question is about volume.

This is the part where most students have difficulty: understanding the problem and knowing what math to apply to solve it. In other words, applying a memorized formula to finding a solution is okay, but only if you first understand the problem.

Memorization is also acceptable when a student is not intellectually ready for the concepts behind certain mathematical operations. In this case, repetition and practice are acceptable means to learning math. I read an article in the *Globe and Mail* written by Margaret Wente, and in it, mathematician and originator of the JUMP program, John Mighton, says that "Kids should be able to think independently, but first you have to function with numbers. Then you layer the concepts on top of that." Sometimes just learning and getting strong with the skills is enough to significantly improve a child's self-confidence and ability. Struggling and struggling to learn a concept that the student is not ready for is only going to lead him or her to feel more defeated. Repetition and practice will work for now.

Habit #8: Being disorganized

How can you hand in your homework on time if you can't find it? How can you review for your test when all of your notes are missing, out of order, or so messy that it is impossible to read them? And how can you finish your homework if you leave your textbook in your locker?

Being disorganized definitely qualifies as a self-destructive habit. Disorganized students are often stressed or panicking because of their own disorganization. Is this you? To find out, think hard about the past month of school and then, as honestly as you can, answer the following questions:

3. How many assignments have you lost or left at home?

4. How often were you late for school because you were at home looking for something?

5. How many times did you get to class only to realize that you had forgotten your pencil or textbook or calculator, or had run out of paper again, or had brought the wrong binder?

6. Did you use your planner on a daily basis?

7. Did you lose your planner?

8. How many tests or quizzes were you unprepared for because you forgot about them?

9. Did you put loose worksheets and homework directly into your backpack so that they could get ripped or crumpled?

10. How often was your bedroom clean?

11. How much stress do you think being disorganized caused you?

12. How would your life have been easier if you had been free of all these organization issues?

If you are disorganized, you probably feel like you spend half of your life one step behind everyone else, not to mention constantly worried and under stress. The great news is, later on in this book you will learn how just a few small changes can make your life amazingly easier.

Habit #9: Having the "let's get this over with" attitude, instead of the "I want to figure this out" attitude

The "let's get this over with" attitude is one in which the student has no inner drive to do well in and understand math. He may want to get a good grade, but he is not willing to put in the effort to do so. For this student, everything is more important than math.

If any of the following consistently applies to you, you may suffer from the "let's get this over with" attitude:

- You race through your homework and then close your books, even if you do not understand the work you did.

- Your homework is messy or incomplete.

- You copy other people's work.

- You skip class.

- You cause problems for the teacher during class by not listening or by distracting your classmates.

- You ask the teacher if the lesson is almost over. Or you ask to leave class to go to the restroom or get your textbook or get a drink of water or do anything else you can think of besides being in math class.

If any of these describes you, remember this:

The path of least resistance always ends up being the path of most resistance.

In other words, if you want to take the easy way now and avoid your responsibilities, you will inevitably to have to deal with a harder life later. It's that simple. If you do not try to learn, and you generally don't care, and you constantly try to take the easy way out, then every year math will get harder, and you will like it less and less. On the other hand, if you radically change your attitude and start trying you will probably begin to really enjoy math.

Habit #10: Missing a class and assuming you are not responsible for the lesson

"But I didn't know we had homework. I wasn't here." I have heard this line all too many times. For some reason, it is a common misconception among students that if they miss a class, they do not need to learn the material that was covered that day or complete the homework that was assigned. This is not true. What is true is that *you are responsible for everything that was covered in class whether you were there or not.*

This is particularly true in math, where there is homework nearly everyday. If you miss a class and decide not to get the notes and assignment from the teacher or from someone who was there, you will only create more stress for yourself.

The first problem is that the next class will probably be hard to understand. Remember, math builds on itself—so what you learn today depends on what you learned yesterday. If you learned nothing yesterday, then today's lesson will be that much more difficult. So get the notes from someone and figure them out, do the homework, and get help if you need it. When you return to school, you will be caught up with the rest of the class and feel much happier than if you had avoided the work.

The second problem with not getting the notes and making up the work is that when it comes time to review for the test, you will not be able to properly review the material that was taught while you were away. And by then, it will likely be too late to learn the material. You may still be able to get notes from a friend, but you probably will not have time to copy them *and* do all the homework while studying for the upcoming test. And your teacher may be unwilling to reteach you something covered a few weeks earlier, when she knows that you have not made an effort to learn it until now.

So there it is—a basic summary of some of the most harmful math habits. The rest of the book is about getting these habits out of your life once and for all and replacing them with habits that produce the results you want. This is not an easy thing to do—good habits are extremely difficult to create—but the beauty is that once a behavior becomes a habit, that behavior becomes almost automatic. You will hardly have to think about studying or doing homework; you will just do it. But first you will have to practice the awesome new habits in the chapters that follow, and then watch your life change before your eyes.

Chapter 2: Healing the Injured Spirit

By this time, some of you may have already stopped believing that you can learn math. Maybe you have convinced yourself that you are terrible at it. Perhaps you are still trying, but because you have come to believe there is no way you will ever do well, you have settled for just being able to pass.

If you are this discouraged, I imagine you dread your tests and feel anxious every time you open your textbook or even think about math. And you probably feel like every math class lasts for an eternity, because you don't understand it.

I know what you are going through. I used to believe that I would never learn math. There were a few years of high school when math was like a foreign language to me, and it seemed my math teachers didn't care. Before Grade 12, I never had a teacher suggest I get a tutor or come in after school for extra help. No one offered suggestions as to how I could improve. No one told me that math *is* hard, and that it wasn't my fault if I had trouble with it. No one told me I wasn't stupid.

Other students your age know what you are going through too. One morning, during the time I was writing this book, I asked one of the students I tutor, "If you were to read a book about how to feel better about math and how to improve at it, what would you want it to say?" I was shocked by her answer. She looked up sadly at me and said, "I know this sounds silly, but I would want it to tell me that I'm not stupid and it's not my fault if I don't understand."

Wow, and this girl is bright! She is in an accelerated program and works so hard, yet she has still somehow been taught by the system to believe that she is "stupid" at math. What a tragedy.

Sadder, I could tell you story after story about students like this. So let's get something straight right now: having trouble in math does not mean you lack intelligence. In fact, having trouble at school doesn't mean you lack intelligence. Some of the most successful and intelligent people in the world struggled immensely in school; some of them never even made it to high school. So if you are trying hard in math and still having trouble, try to smile and not worry too much, because ...

1. **You are *not* stupid**, no matter how much you struggle in math.

2. **It is *not* your fault.** Math is hard, and some people are more gifted at it than others. Don't blame or get upset with yourself if you don't understand it, and don't let anyone else get upset with you for struggling with it either.

3. **Where there's a will, there's a way.** You can always do better. Like anything in life, if you practice often and try your best, you will improve.

So believe in yourself and always remember: you are smart. Did you get that? *You are smart!* I have seen so many students lose enormous amounts of self-esteem over math, but the fact is that most people have trouble in math—even adults. It's actually quite normal!

So next time you have difficulty understanding your homework, instead of telling yourself you can't do it, tell yourself that math is just hard and that you are normal.

When It Seems Like Almost Everyone is Better Than You at Math
When I was in high school, I used to get so upset when a friend of mine did well in math—which was almost always—because she hardly had to try at it. Then there was me. I worked hard on my math but rarely did well. I also remember so many times looking at my grade posted on the classroom wall and feeling awful because of all the people who were above me on the list.

But I learned something from these experiences. I learned that some people are just good at math—that there will always be those students who sleep through half the year and still get A's. I also learned not to worry about it because, the fact is, everyone is different, and comparing yourself with others is only self-destructive and negative, and will not help you improve. In life, there will always be someone better than you at something, but you will shine in other areas. Everyone has a gift; for some, that gift is math, and for others, that gift is something else. My gift is helping people learn to understand math and enjoy it. Yours might be landscaping or dancing or woodworking or playing drums or understanding people. But the point is that *not being gifted in math does not mean you can't be successful at it.*

So work hard and try your best. And before you get upset at your best friend or sibling (or whoever) who happens to be amazing at math, remember to be grateful for the experience because of what you are learning from it. After all, you are learning how to work hard for what you want, be self-disciplined, have a positive attitude, and be resourceful. These lessons will take you far in life. And when you graduate, these experiences will help you feel that much more proud of what you have accomplished.

Don't Worry—Things Will Get Better

I know what it feels like to want to give up. There were so many times that I honestly did not believe that I would make it through high school math and just wanted to quit. If this is how you feel, don't worry—there is hope!

But we can only get you through this if you are ready and willing to try some new things. In life, if you want different results, you have to take different actions. After all, if what you are doing now were working, you would not be reading this, right?

So let's get you motivated and excited again! Let's create a drive so strong in you that nothing will be able to stop you! To do this, all you need to do is sit back and imagine with me...

Imagine you are finished reading this book. You have followed my suggestions and created new math habits, and your life is beginning to change. The frustration, anxiety, fear, and other bad feelings have begun to transform into confidence, curiosity, patience, and even joy.

Imagine coming home with amazingly better test scores. Imagine getting 94% on a math test, or 98%, or even 100%, and not because you were lucky but because you really knew the math. What would it feel like to look at a math test and know how to do all of the questions? Wouldn't that be awesome?

And how would you feel if you understood an entire math lesson? Imagine answering the teacher's questions and even helping your friends. Would you still hate math? I doubt it. (The truth is that most people don't hate math at all—what they hate is the fact that it never made any sense to them.)

Imagine being in a position where you finally understand math, and actually enjoy it.

All of this is possible for you.

In fact, I am certain that after you begin using even half of the strategies in this book, you will do and feel better. So many wonderful things will start to happen for you, and you won't have to be frustrated, scared, worried, or confused anymore. You will stop doubting yourself and feeling bad about yourself and will at least start to believe that you *can* do math.

Are you excited yet? I thought so! So let's get started! Here are my first three suggestions for you. Let's call these our "feel better" strategies.

- **Don't take it personally.** I know how easy it is to get discouraged and upset about math, but it is crucial not to let this affect your self-esteem. Anytime you feel self-doubt or negative thoughts coming on, push them out of your mind and replace them with something positive. If you are having a tough time understanding your homework, for example, keep telling yourself, "I just need to get a little help on this" or "Next time I will ask more questions during class." Never ever say "I can't do this" or "I suck at math." Put all of these types of thoughts on the swearwords list—they will only make you feel worse and will not improve your situation. And besides, *no one sucks at math until they decide they do.* (Note: In later chapters I will show you how to create a plan to help you work out whatever negative thoughts or feelings are upsetting you, so stay tuned.)

- **Stop procrastinating.** Agonizing over doing your math homework is usually far more stressful than actually doing it. Trust me— avoiding my homework at all costs used to be one of my favorite pastimes. Yet the more I would try to avoid it, the more anxiety I would create for myself. It is easy to convince yourself to do it later, but the problem is that when later comes, you most likely will still not want to do it. To avoid this situation, try starting your homework right after school. And the next time you really do not want to do it, think, "All I have to do is open the book and do one question." After trying one question, you may find that the work is not as bad as you had imagined. Just start. If you are afraid that you won't know how to do the questions, open the book anyway. Then after following the earlier homework suggestions in this book—reading your notes and the textbook—just try one question and see. I bet you will be fine— and if not, don't panic! Leave the questions you are struggling with and get help on them later.

  Be careful though, not to spend too much time on a question that you really will not be able to figure out without help. This is damaging to your emotional state, and will probably make you want to procrastinate even more the next time. After all, how can you possibly stay positive about math if you work yourself into deeper and deeper frustration every time you do it? Remember, it's just a math question.

  And if you start talking yourself into watching television, playing video games, or talking on the phone instead of doing your homework, remember that being lazy will actually be harder on you than doing the work. Sound crazy? Think about it—the emotional pain you cause yourself by avoiding the work and then doing poorly on tests or not getting your homework finished on time is far worse than the pain you may have to endure to do homework and study a little every night.

  If you do manage to sit down and start only to find yourself staring blankly at your open book, then maybe it's time to talk to your parents or teacher. And be honest about it—tell them you can't seem to get motivated. They might be able to help you come up with a plan. Another option is to find a classmate to do your homework with; you can coach and motivate each other.

- **What about boredom?** "But math is boring; I don't want to do it." How many times have you said that in your life? I must admit, I agree—math can be boring. And most textbooks are arranged in such a way that the math is sometimes dreadfully repetitive, giving the exact same kind of question over and over again.

So try this on your homework: don't do the questions in order. That is, definitely do a few of the easier ones first so you understand the basics, but then start mixing them up. Do #1 and then #5 and then #7 and then #2, and so on. This method is more challenging and interesting, and, as a bonus, more similar to what you might see on a test. It is a method that will also help you to think about the math, because you will not be able to answer a question simply by doing what you did on the question before; instead, you will have to recognize what type of question it is and then figure out how to solve it.

You may feel unsure about the method at first, but give it some time. If you consistently do your homework questions out of order, you may find yourself becoming comfortable with it, not to mention calmer during tests.

Keep in mind, as well, that this method of doing homework is not common, so if you do try it, let your teacher know what you are doing and why. This will prevent any misunderstandings or lost homework marks later.

NOW STOP EVERYTHING!

Reread the "feel better" strategies above and then put this book down. Do not read past this section until you have mastered these strategies. Mastery means making the strategies a part of you, letting them become so ingrained in your mind that you apply them consistently without even thinking about it. So give yourself a few weeks to practice using these strategies every single day, every chance you get, until they begin to feel natural.

I want you to do this because reading alone is not enough to create results; you must also take action! For example, let's say you want to learn how to drive. Reading about how to drive and watching other people drive will only teach you about 20% of what you need to know.

The rest is practice—driving and driving and driving some more, until it starts to get easier and then eventually becomes so natural that you hardly have to think about it.

Trust me, making these three small changes (using the "feel better" strategies) will be worth the effort. Use them as often as you can for the next month and then come back and pick up the book again. In that short amount of time, I bet you will have completely changed your outlook—not only toward math but also on your own ability to change. Good luck!

Did you do it? Did you master the "feel better" strategies? If not, go back and keep trying! You are not ready to move forward yet.

If you really tried to incorporate the strategies into your daily life, congratulations! You have made a huge amount of progress. What was the experience like? How do you feel? If you managed to turn these strategies into habits, I imagine that you feel much more confident and less stressed than before. So now it is time to move forward ... well, almost time. There is one final thing you need to do: convince yourself that making major changes in your math habits is *absolutely necessary;* because unless you genuinely convince yourself of that, you will not be able to follow through with the strategies in this book (including the ones you have already been working with).

So convince yourself that you have had enough! You have had enough of not understanding and of feeling bad about yourself, and you will not tolerate even one more low grade! This is it. You are done. Fed up. *Enough!*

Unless you are sincere in your desire to make a lasting change, your motivation will dwindle.

Chapter 3: The Power of Persistence

Persistence is deciding on and committing to a course of action, and then not giving up no matter what. You have already taken the first step toward developing persistence—you have made the honest decision that you want to try, that you have had enough of poor results. You are ready. This is the first step to maintaining a course of action because without a burning desire to succeed, the smallest obstacles and challenges will make you want to give up.

If you are not convinced, think of all the people you know who are persistent and you will notice they all have something in common—*they want something and they want it badly*. Persistence and dedication come naturally from that place of unshakable desire, and it is only when you get to this point that you will refuse to let anything stop you from reaching your goal.

Your next step is to develop a plan of action that will give you the results you want, and then follow through with it. So find out what it takes to do well, and then build a plan around it. There are numerous suggestions throughout this book that will help you do that. When coming up with your action plan, keep in mind that success in anything in life requires *sustained focused effort,* so also make sure your plan is realistic for *you* so that you can stick with it. That does not mean that your plan should not be challenging, but only that it should be made with consideration of factors such as your existing schedule and workload. Once you have developed your plan, break it down into a series of good practices and then commit to those practices. This will become easier as you get used to them. If you do not make this commitment, nothing will change because in math (and in anything else, for that matter) if you work hard for a week and then lose focus for a week and they try again for a few days only to lose interest again, you will not get the results you want. Tennis champions and marathon runners hold long training sessions at least six days a week. Models have diets, trainers, and strict workout schedules. Pianists practice for hours upon end. These dedicated people have probably overcome enormous obstacles to get where they are, and I doubt they make excuses or feel sorry for themselves—they just do what they have to do to succeed.

This leads me to my next point. After developing an action plan and learning to carry it out on a daily basis, your third step in persistence is learning how to maintain your desire. We have to keep you wanting this for the rest of high school, if not longer—not just for the week or the month.

It is common for all of us to get inspired and change our behavior for a while, only to gradually fall back into our old ways. Consider all the people you know who have made New Year's resolutions; after only a month, *maybe* 5% of them are still committed. This is why gyms and fitness centers are always busiest in January, but by the end of February have usually slowed down again.

At times on your new journey, math will seriously test your spirit. You will, without a doubt, have experiences that will try your commitment, tempt you to doubt yourself, and make you want to quit—like failing a test that you studied really hard for or not being able to find the motivation to do your homework.

Unfortunately, sometimes the desire that motivates you to change in the first place is not enough to carry you through these hard times. You have to constantly renew your desire. Here are a couple of ways to do this:

Stay positive. When you encounter negative experiences, you will likely feel frustrated, angry, or upset, which will lead you to think negatively—*but that thinking only makes you feel worse*. Do not let such negative thoughts stay in your mind, because they will cause you to lose your confidence and focus. Try as hard as you can to keep your spirits up (you will learn more about how to do that in the next chapter).

Focus on solutions. In the next chapter, I will show you how to train yourself so that, when you start to feel discouraged or upset, you can shift gears in such a way that the first thought that enters your mind is "What do I need to do in order to feel better?" And then do it. If you get stuck, get help. If you fail a test, do the corrections and then become determined to study harder or better next time. If you don't like your teacher, find a tutor and work hard anyway. Do whatever it takes to do well.

Don't complain. If you tend to complain, try to break the habit. Venting once in awhile is healthy, but complaining regularly is negative because it involves focusing on the problem instead of a solution. Even if complaining feels good for the moment, in the long run it only drains your energy and puts the blame on someone else.

Finally, maybe the best way to stay motivated is this:

Whenever you feel like giving up, remember your previous pain.

While this may seem like negative thinking, it actually is not. Instead, it is a way to acknowledge that your efforts are what have gotten you out of—and what can continue to keep you out of—that pain. Remember how awful it feels to fail a test or to bring home a report card with a C- or even an F in math. Remember how frustrating it is not to understand your homework. Remember how bad it feels coming to class without your assignment done. Remember what it feels like to be disappointed in yourself. If you regularly remind yourself of the emotional pain you went through before you began changing your habits, you will be more likely to want to stay on your current upward path.

Also remember how hard you have worked to get to where you are now and how happy you felt overcoming previous challenges. This is not the "happy" you feel when you have a fun weekend or watch an awesome movie, but a much deeper, more meaningful kind of happy—belief in yourself and pride in knowing you have accomplished an amazing goal. That is a happiness that no one can take from you.

So stay positive, get help when you need it, and don't *ever* question your intelligence. And instead of complaining and feeling sorry for yourself, use your energy to get the problem solved, even if that means calling someone for help. This is persistence—*continuing* to try your best and to maintain a positive attitude in the face of any challenge.

25

Chapter 4: Change Your Thoughts to Change Your Grade

As mentioned in the previous chapter, one of the best ways to ensure that you remain persistent with your math goals is to think positively and focus on solutions. While this is true, I know from experience that it is also easier said than done. High school math can be extremely challenging, even after you have committed to trying harder and creating better habits. Even in my senior year of high school and at university, when I was trying my hardest, I would fail the occasional test or not understand my homework.

It is during these frustrating and discouraging times when it is the most difficult, but also the most important, to stay positive. In this chapter, I will show you how to manage and transform negative emotions, so that you can stay focused and confident. The chapter is based on the following powerful idea:

> *The thoughts and beliefs you have about math, and the actions you take because of those thoughts, will directly influence how well you do.*

Read that again because it is a surprising truth, and it means that you are in control of how well you do. Can you believe that the thoughts and beliefs you have about math will affect your results probably more than any other factor? In fact, because our thoughts drive our actions, your beliefs about anything will directly affect your results in that area. Everything we do begins in our mind. And guess what? Most of our thoughts are influenced by what we are feeling. So feelings influence the thoughts that create the actions. The process goes like this:

$$feelings \rightarrow thoughts \rightarrow action$$

It makes perfect sense when you think about it. When people feel hungry (feeling), they think about what food they might like (thought), and then they eat (action). When they feel lonely (feeling), they consider calling a friend (thought), and then they pick up the phone (action). When they feel motivated and full of energy, they might go for a run; when they feel tired, they might sleep; and when *they feel frustrated with math, they might think it is pointless and give up.*

But imagine if a student stopped herself just after she had the thought that math was pointless, recognized how negative the thought was, and tried to replace it with something better. That process would look like this.

$$feelings \rightarrow thoughts \rightarrow stop \ and \ think \rightarrow more \ positive \ thoughts \rightarrow$$
$$more \ positive \ action$$

Now let's back up a bit. Did you even know you had feelings about math?

If you think about it you will probably find that you experience some sort of emotion about math daily. One day, you might be overjoyed about doing well on a test, and then a few days later, go home from school upset because you didn't understand the math lesson at all.

All of these feelings influence your thoughts about math, which can be dangerous when those feelings are often negative. Without realizing it, students can convince themselves of all kinds of awful things about their ability or their intelligence, and those thoughts affect their future performance in math, which starts the cycle all over again—more negative performances create more negative feelings and thoughts, and so on.

Since your feelings are so intertwined with how well you do in math, the question becomes: Can you change your feelings and use them to your advantage? The answer is yes!

As mentioned above, you need to interrupt the *feeling → thought → action* process between thought and action, and create a better thought that will help you feel more positive and therefore motivate you to positive action; I call this new thought a **positive-action thought**. A positive-action thought is a thought you create that will inspire you to take positive action.

So now the process might look more like this:

feelings → thoughts → stop and think (check your thoughts; are they negative?) → (if negative, change them to positive-action thoughts) more positive thoughts → positive feelings → constructive action

To see how this might work, let's look at an example. Jack and Jill just got their math tests back. Both students struggle in the subject. Jack received 48% on his test, and Jill got 46%. Jack is a happy and easy-going fellow and always tries to look on the bright side. Jill has accepted the idea that she is terrible at math; she tends to panic on tests, and every time she gets a test back, she worries about her future. Now compare Jack's reaction below to Jill's:

Jack:
Feelings: disappointment, frustration
Thoughts (positive-action): This feels really awful. I wish I'd studied more. But maybe I can go over it and figure out my mistakes. Then at least I'll know where I went wrong.
Positive feelings: motivation, determination to learn from the experience and not to let himself feel so awful again from "bombing" a test.
Constructive action: Studies harder for the next test, and does not procrastinate, so that if he has trouble, he has time to get help. Does better on the next test.

Jill:
Feelings: inadequacy, helplessness, frustration, anger, sadness
Thoughts: I hate math. It doesn't matter how hard I try, I can't do it. Why do I even bother? I'm never going to be good at it.
Action: Hides the test from her parents, does not get help, has the same negative experience on the next test.

Between Jack and Jill, who do think improved in math? Can you see now how important it is to create positive thoughts? It's crucial!

If you are not careful about how you handle your emotions, they can drive you to giving up hope. On the bright side, you have the ability to change your thoughts and create an emotion that will motivate you like crazy! Isn't that awesome?

The first step is to *train* yourself to recognize whether your thoughts are constructive or destructive (positive or negative). Ask yourself, "What are my exact thoughts right now? Will these thoughts help me or hurt me?" If they are destructive, then ask yourself, "How can I think differently right now so that I don't let this feeling take over and make me do something that will make my situation worse?"

This will be hard at first, especially when you feel really down. But with practice, your emotions *will* begin to change, and you will start to feel better.

So from now on, watch your emotions in all areas of your life—not just math— and practice not letting them control you. You choose. When you feel sad or angry, you do not have to stay in those feelings. You can take action and use those feelings to make a change for the better. Use your emotions as fuel for change!

Now let's see how this more positive approach would have worked in Jill's situation.

Jill:

Feelings: inadequacy, helplessness, frustration, anger, sadness

Thoughts (positive-action): Why did I fail that test? What could I have done better? I know I am capable of doing well in math. It's hard for me, but that's okay. Maybe if I sit down and talk with my teacher, we can figure out what I am doing wrong, and I can fix it. Maybe I can talk to my parents about getting a tutor, too.

Action: Talks to her teacher and her parents, finds out where she needs help, and works harder than ever before, both in class and on studying for her next test. Does much better on her next test.

Wow, what a difference!

Learn to use your emotions to help you make good decisions instead of ones that will make your situation worse and you will be amazed at what you can accomplish. In fact if you become really good at creating positive thoughts, you can be less "smart" than other students at math and still do extremely well! It is all about the choices you make—you are in control.

Chapter 5: Slam the Door on Test Anxiety

Test anxiety is the all-too-common student experience of panicking during a test and then not being able to think clearly. It usually happens something like this: The teacher hands out the test; you look at the first few questions and find that nothing really looks familiar. Even the questions you have practiced seem tricky. You start to feel horribly nervous. Your stomach tightens and you begin to feel hot. Suddenly you have no faith in what you thought you knew, and you panic and start imagining the worst. Before long, you are so overwhelmed with stress and worry that you have little hope of concentrating on the test.

Sound familiar? If so, you are not alone. I hear students talk about this experience all the time. They say things like, "I don't know what happened. I just froze!" or "My mind went blank." or "I couldn't think."

The Symptoms of Test Anxiety

Before we discuss the causes of test anxiety, I want to take a closer look at what actually happens to the student who experiences it. After reading the description above and thinking about your own experience, you may notice that test anxiety has three distinct components: physical, psychological, and emotional. And as shown in Diagram 1 below, all three of these components feed one another.

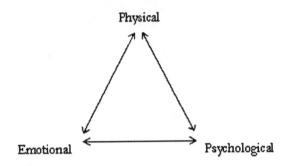

The physical part of test anxiety is usually experienced as an increased heart rate, an upset stomach or "butterflies in the stomach," a feeling of being hot or as if you are breaking out in a sweat, and even dizziness. The emotional component involves feelings of fear, frustration, sadness, anger, or any combination of these. Finally, the psychological component is the negative self-talk: "I'm going to fail this course if I don't do well on this test." or "I am so bad at math—I can't believe I don't know any of this." or "My parents are going to kill me; I can't fail this test."

This cycle of physical, emotional, and psychological stress is extremely distracting and any student experiencing it will have incredible difficulty focusing. That is why test anxiety is such a problem—it disables the student's thinking process. In her book, *Overcoming Math Anxiety*, Sheila Tobias said it perfectly: "As long as a student is burdened by negative feelings, she cannot concentrate on the mathematics. And concentration is the key." (244)

The good news: if you are aware of what is happening to you when you experience these thoughts, feelings, and physical symptoms, you will be better able to interrupt the test-anxiety process and refocus.

An even better solution is to do everything you can to avoid having the anxiety in the first place. To do that, we need to begin by exploring the causes of test anxiety.

Determining the Cause of Your Anxiety

Below, I have listed three main causes of test anxiety. Ask yourself whether any of these apply to you, and if they do, highlight or underline them, and come back to this section of the book later to see if you have made some progress.

1. **Lack of preparation.** The factor that seems to make students most anxious during tests is a lack of preparation. In fact, *most times, test anxiety is a direct result of not being prepared for the test.*

   When the students I tutor write math tests at school, they are sometimes able to bring the tests home afterwards. I usually look over these tests with the students to see where they made mistakes. Most students who do not do well tell me that they "froze" on the test. What I have found repeatedly with these students, is while they may have lost a few marks from being nervous, most of their incorrect answers were on problems that they simply did not know how to do. I know this because when I ask these students to redo the questions they "froze" on, most times, they cannot do it, even when they are with me in their homes and not in a test environment.

   On the other hand, it is rare for students who know the material well to totally freeze on a test.

   So if you struggle with test anxiety, the first thing to do is look at how well and how much you are studying. This is not to say that being well-prepared will guarantee that you not feel test anxiety. You may still feel it, but your confidence and strength from all the preparation will likely override the anxiety. It will be easier to trust what you know and stay calm even while you feel anxious.

2. **Tests and quizzes tend to have a different format from homework.** A typical math lesson tends to involve only one or two new concepts, so on any given homework assignment, the student is applying the same ideas over and over again. In addition, homework questions are usually grouped together in such a way that the questions in each group are almost identical. So other than word problems, homework offers very little opportunity to practice creative thinking skills.

   Tests, on the other hand, cover material from several different lessons, and the questions are not always grouped according to concept, so students must decide how to solve each problem by *thinking.* Students are forced to be creative and recognize what type of question they are dealing with, as well as how to solve it.

Unfortunately, textbooks and math classes rarely offer the kind of practice necessary to enable students to look at a random math question on a test and be able to come up with or recall the method for solving it. And of course, this process becomes even more difficult when a student is in a state of high stress.

3. **Pressure to succeed and the fear of failure.** Some students have tremendous pressure from home to do well in math. Others may need a certain grade in math to be accepted into the university of their choice.

   The classroom environment itself can also induce test anxiety. Sometimes the atmosphere in a classroom can become extremely tense during test time. The silence in the room combined with your own nervousness, can be overwhelming and very distracting. It is no wonder students have difficulty concentrating.

To learn more about your own test anxiety (if it is a common experience for you), find a test that you "froze" on and take an honest look through it, one question at a time, to see if you can discover what went wrong. Were you genuinely anxious or was it a matter of not understanding the math? Would more practice have helped? Were you distracted by your thoughts? Once you have determined the cause of your anxiety, you are ready for the next step.

Managing and Reducing Your Anxiety
Now that we have explored the most common causes of test anxiety, it is time to focus on solutions. Here are some things you can do that may help you.

Before the Test

- **Make sure you understand all of the material in the chapter.** Of course you will get anxious during your test if you do not know the math! So before your test, begin my looking over your notes from that chapter. Then review all of the corresponding homework and try some of the questions again. If there were any questions that you had trouble with, make sure you learn how to do them now—those are usually the ones that show up on tests. Also, look through your quizzes from the unit and make sure you understand any mistakes you made. Then try to do the problems on the quizzes again, without looking at the answers.

• **Create practice tests and quizzes to prepare for the real thing.** *This may be the single most effective thing you can do to improve your test results.* Doing a practice test gives you the opportunity to see how you are doing with the concepts in a unit and what areas need more work.

The greatest thing about practice tests is that they give you the opportunity to make mistakes without losing marks! You can then learn from those mistakes, so that when it does count, you will get the questions right.

Another benefit to writing practice tests is if you write them on a regular basis, you will likely become more relaxed in a test environment. This process of exposing yourself regularly to what would normally induce fear, is called desensitization. I have tutored students who in the past had been terrified during math tests, but after having so much practice creating and doing their own quizzes, they became accustomed to the pressure so tests no longer scared them. Several of these students have come back to me expressing feelings of joy and relief because they found themselves calm and able to think clearly during a test or quiz. What a wonderful thing!

How to do it: You can ask a classmate or your parents to make quizzes for you, or even make them yourself. Your teacher would probably be happy to give you old quizzes to practice with, as well. Make sure that whoever makes the quiz has the correct answers, so that you can mark your work. Once you have marked the quiz, focus on learning from the questions that you did incorrectly. Then do another practice quiz like this, and then another. Keep doing them until you are consistently getting all of the questions right. If you can get above 90% on five quizzes in a row, you are probably ready for your test.

During the Test

- **Focus on the test.** Imagine that you are driving home from a friend's house late one night. He lives quite far, so you decide to take the freeway. As you are driving, you suddenly find yourself in a frighteningly thick fog. It is so thick that you can hardly see beyond five feet! You turn on your high beams, but that makes it worse. What do you do? Panic is an understandable reaction, but if you worry about the fog, how will you be able to concentrate on the road? If you are focused only on how afraid you are, you may not be able to stay calm enough to drive safely and avoid an accident. Wouldn't it be better to at least try to forget about being afraid and instead focus on not crashing? Clear your head and focus completely on driving, and you will be more likely to make the right decisions.

  The same is true of test anxiety; it is a form of fear, and you have the choice whether to focus on that fear or focus on the test. The best way to eliminate the anxiety is to think about the math. Focus on your test. Stay present.

  One way to do this is to take on only one question at a time. Tell yourself something like, "Right now, I only need to do this one question. I can put my worry aside just for a minute, so I can do this question. This is the only thing in the world right now." Then when you're finished with that one, do the same thing for the next question, and the next.

  Keeping in mind that fear is an emotion, another thing you can do is to use the ideas presented in Chapter 4 about how to transform your fear into positive action. If you remember, the process looks like this:

feelings → thoughts → stop and think (check your thoughts; are they negative?) → (if negative, change them to positive-action thoughts) more positive thoughts → positive feelings → constructive action

  Otherwise, you may end up focusing on negative thoughts such as, "This is so hard. I don't know anything. I'm going to fail!" *Thoughts like these will distract you,* and you will become even more emotional and less able to focus. As Tobias observed,

"Freezing and quitting may be as much the result of destructive self-talk as of unfamiliarity with the problem." (69)

Intercept those negative thoughts with positive-action thoughts, such as, "I may be anxious and scared right now, but that's okay. There is nothing wrong with feeling scared. I can still focus on my test. I just have to try extra hard to understand what the questions are asking." And then focus fully and completely on the math. As Tobias goes on to say, "… if we can talk ourselves into feeling comfortable and secure, we may let in a good idea." (69)

You have the power to think yourself out of anxiety and to succeed at math.

- **Do the easy ones first.** Often on tests, some of the harder questions are at the beginning; this means there is a good chance you may get confused or stuck before you even really start. If this happens to you, try not to panic or spend too much time on those questions—you can go back to them later. The best strategy when writing a test, however, is to go through it once and do the questions that you think are fairly easy. The process of doing those questions will calm you down and give you confidence, so that you can move onto increasingly challenging problems.

- **Be patient when solving word problems.** In general, word problems seem to be the area where students struggle most. In a test environment, they can be even more intimidating. Again, try not to dwell on your anxiety—this will only distract you. Many students get so nervous that they cannot stay focused long enough even to understand what the problem is about, let alone solve it.

  In the following chapter, I will introduce you to a strategy you can always use to solve word problems. Once you are comfortable with using the strategy, you may find that word problems will not bother you as much. This is mainly because this strategy gives you a place to start every time, as well as a series of steps that guide you to the final solution.

Chapter 6: How to Master Word Problems

Time and time again, I have heard students tell me how much they "hate" word problems. What I noticed is the students who dislike word problems so much are also the ones who struggle with them. This is the case with several of my students, and after much observation, I began to see the problem—these students do not have a strategy for solving word problems, so when they encounter a word problem, they do not know how to begin or even how to read the problem correctly. And if they do attempt the problem, they seem to immediately begin searching for the answer or the exact steps to get to it, rather than trying to think the problem through. As a result, most times the students become overwhelmed and frustrated, until finally they give up.

On the contrary, students who read word problems carefully and try to think them through, tend to enjoy them more and be successful with them.

If you struggle with word problems, then the strategy that follows should be a very helpful tool for you. The great thing about this strategy is that it always gives you a place to start, and then a process to help you through to the final solution.

Once you have had some practice working with this strategy, and you know how to read and interpret word problems, you will find that most times they are not as difficult as they first appear to be; understanding word problems is just a matter of being patient and trying to understand each sentence, one at a time. Be prepared though because this strategy does not guarantee that you will never get stuck again. Getting stuck on word problems is a natural and positive part of learning math. So learn to be patient with yourself, always remembering that the struggle is part of the process.

The Strategy

Step 1: Put your pencil down and read the question

Yes, that's right, put it down! Students often read the problem so quickly (if at all) that they do not really understand what it says. They try to rush to the "math part," writing and calculating before they have even decided what they are trying to figure out. This is like leaving your house in the morning without knowing where you are going or how to get there—seems pointless, don't you think?

So before you start writing or doing any calculations, read the question through a few times, and try to understand what it is about. Make sure you know what all of the words mean, and if you are unsure of any, look them up in the glossary at the back of your textbook or on the Internet. I know this is not possible during a test, but hopefully by test time you will be familiar with all of the vocabulary from the unit. The idea is to find out what the question is about and what you are being asked to find. Once you understand these two things, you will find that suddenly you are able to actually think about how to solve the problem. And after some practice, you will begin to gain a feeling of independence knowing that you can use your own thinking skills to come up with a solution.

It is also helpful at this time to decide what your answer should look like, so that when you arrive at one, you will know whether it seems reasonable.

This step is all about understanding—searching for meaning. Remember, with word problems performing the actual calculations is often one of the last steps. Thinking and planning have to come first.

Step 2: Draw a diagram, if possible

Once you know what is going on in the question, you may want to draw and label a diagram. This is not necessary for all word problems, but if the question involves geometrical items such as triangles, perimeter, or area, or gives visual pictures such as rectangular gardens, pools, sidewalks, or picture frames, a diagram will usually help. Once you draw it, you can also label it, and the question will start to make a little more sense.

Step 3: Come up with a plan

Now that you know what the question is about and what type of answer you are looking for, it is time to come up with the strategy to solve the problem. This step involves figuring out what math you will need and, when necessary, setting up the appropriate equations. If you get stuck here, keep trying. If after a while, you still cannot see what to do, look back in your notes or get a hint from a friend or your teacher. But try not to get the full solution—a hint may be enough for you to finish it on your own. If you are writing a test and get stuck at this stage, leave the problem and come back to it later.

Step 4: Carry out your plan

Do the math!

Step 5: Reread the question and make sure you answered what was asked

This is the step that students often forget, especially if the student did not read the question carefully to begin with. Once you think you are finished a word problem, reread the question to make sure you have actually answered it. For example, if the question asks for the amount of money in a collection of quarters, and your answer is $x = 2$, you are not finished yet, as $x = 2$ is not an amount of money. The acceptable answer would be $2. Writing a sentence answer is even better because it forces you to look at the question again.

Step 6: Check to see that your answer is reasonable

The last step is to check whether or not your answer makes sense. For example, if the problem asks for the ages of two brothers, and you get an answer of 166 and 290, you have probably made a mistake (or else these two brothers have broken all kinds of world records!). Or say the question asks for the number of tickets sold to a school play, and you get a decimal answer like 0.8; you know that you have made a mistake because it is impossible to sell 0.8 tickets. It is surprising how many students hand in tests and homework with totally unreasonable answers, when all it takes to avoid this is a quick check.

Now, let me take you through these six steps using an actual word problem, so you can see how the method works. Then you try!

> A 10-meter ladder is leaning against the side of a building. The bottom of the ladder is 3 meters from the base of the wall. How far up the wall is the top of the ladder, to the nearest tenth of a meter?

Step 1: Put your pencil down and read the question

Read the question a few times slowly. Do you know what all the words mean? Do you know what you are trying to find out? The information we have is that the ladder is 10 meters long and is leaning against the side of the building, and that the bottom of the ladder is 3 meters from the base of the building. I know that the ground will form a right angle with the side of the building. I see now that this is a right-triangle question, and the length of the ladder is the hypotenuse. After another read through, I also know what the question is asking me: to determine the distance from the ground to where the ladder is leaning against the building. Now I can move on to the next step.

Step 2: Draw a diagram, if possible

I know 10 meters is the hypotenuse, because when I draw the ladder leaning against a wall, the ladder is the longest side of a right triangle. I also know that 3 meters is the length of the line along the bottom of the triangle (the ground) because it represents the distance from the *bottom* of the ladder to the *base* of the building. I don't know how long the other side of the triangle is, and that is what I need to find. Since this side will be the height of the triangle, I can call that measurement h (for height).

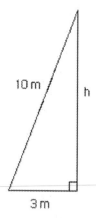

Step 3: Come up with a plan

Since I am dealing with the missing side length of a right triangle, I will need to use the Pythagorean Theorem ($a^2 + b^2 = c^2$) to solve this.

Step 4: Carry out your plan

$3^2 + h^2 = 10^2$ (since 10 is the hypotenuse)

$9 + h^2 = 100$

$h^2 = 100 - 9$ (subtract 9 from both sides)

$h^2 = 91$

$h = \sqrt{91} = 9.539$

Step 5: Reread the question and make sure you answered what was asked

The question tells me to find out how far up the wall the top of the ladder is to the nearest tenth of a meter. *I missed two things*: units and rounding. The nearest tenth of a meter means one decimal place, and the units are meters, so my final answer should be 9.5 meters.

Step 6: Check if your answer is reasonable

Since the ladder is 10 meters long, 9.5 meters sounds like a reasonable height for the ladder to be up the wall. (Note: I found this problem on a worksheet that was given to a class of grade 9 students. Do you notice anything strange about the problem? Hint: when was the last time you saw a ladder that was 10 meters long?)

Do you see how well this strategy works? It is a great tool to have because you can use it for any word problem. And because it gives you at least somewhere to start every time, you will be less likely to feel stuck right away.

Try it out for yourself right now. Find two or three word problems in your math textbook and practice the steps. At first, it may seem to take longer than the way you normally solve them, and it may even seem more difficult, but once you get used to using the strategy you will probably find that it is actually much easier. It is more systematic; it gives you a procedure to follow each time, so you always have somewhere to start, and so you don't accidentally miss any steps. Your word-problem anxiety will be a thing of the past!

The Mysterious "Aha" Moment

There is an interesting and mysterious thing that happens when you solve a word problem; you reach a point where you suddenly "see" the solution—the light goes on, and, all at once, you get it.

Sometimes, you may see the solution right away. Other times, you may not see it at all. And still other times, you may make some progress, but then not know where to go next; you might think of similar problems that you have seen or formulas that could apply as you try to decide what math will take you to the solution. If you have the tools you need to solve the problem, eventually—aha!—you will usually see what to do.

Regardless of what happens, the idea is to engage in a kind-of mental exploration, thinking of several possibilities, so that you can eventually "see" the pathway to the solution. The process is not so much mathematical as it is creative. As Tobias puts it, "Obviously, problem solving is not really a matter of making logical deductions from memorized formulas, but an exercise in imagination." (142)

In this sense, word problems are quite similar to riddles. Once we understand what the riddle is about, we imagine of all kinds of possibilities and then finally (if we're lucky) the answer just pops into our head. Consider this riddle, for example:

> A black dog is sleeping in the middle of a black road that has no streetlights, and there is no moonlight. A car comes quickly down

the road with its lights off and manages to steer around the dog. How did the driver know the dog was there?

Answer: The driver saw the dog from a block away. It was daytime.

Coming up with this answer involves a creative process. It involves thinking outside the box, and it is only through using the imagination that the solution appears.

If you have tried and tried to solve a word problem and still cannot see what to do, then the best strategy is to get help. Spending too much time on a problem and getting frustrated is actually unproductive; it is difficult to remain in a curious and creative state of mind when you are angry or frustrated. So get help, and when you are getting help, try to understand the solution and remember it, so that next time you see a similar problem, you will know what to do. The more problems you learn well (with or without help), the easier and more natural problem-solving will become.

Chapter 7: New Habits That Will Turn Your Life Around (In Math, of Course)

You have come so far in learning how to transform your and results in math. In this chapter, I will introduce you to even more awesome habits that will help you along the way. If you are not already getting amazingly better grades and feeling better about yourself, then it may be time for you to learn these new habits and make them your own. Remember, the key to success in anything in life is creating habits that deliver the results you want.

1. **Create a positive study environment that works for *you*.**
   Do not think there is something wrong with you if you have trouble concentrating in a quiet room by yourself. In high school, I often studied like this because I thought it was the right thing to do, but I was miserable, and I think now I understand why. For most teenagers, it is not natural to sit alone in a quiet room and study. Most adolescents that I have worked with find that environment to be "too serious and stressful." If you agree, here are a few alternatives you might consider.

   - Listen to quiet music. Music is not overly distracting for most people, provided it is not too loud. It can actually be quite soothing and help to reduce stress.

   - Find someone to study with—a partner or even a group. Do homework and study together as much as possible.

Support one another emotionally; coach one another. This means giving pep talks and cheering one another on, when necessary; it means encouraging one another; it means having high standards for yourself and everyone else in the group. The teenage years are usually the most social time in a person's life, so allow yourself to learn in a social environment if that is how you are comfortable.

• If staying focused and productive in a study group proves to be too difficult for you, or if the social aspect does not appeal to you, working with a tutor is another option.

It is up to each student to find out what works best for them. But remember, math does not have to be lonely and boring!

2. Make math a priority.

Improving in math is not easy. It requires you to focus, to push yourself, and sometimes to put it before other things that are important to you. I have had several students who have done poorly in math because they had too many obligations, or else simply did not feel like putting much time into studying and doing homework.

Unfortunately, it is difficult to have the best of both worlds. Most students who improve in math find that they have to make sacrifices along the way, including spending less time on sports or other activities. So sit down with a family member or friend, or even by yourself, and figure out what is most important to you right now, and whether you have time in your schedule to spend extra time on math. If you do not have the time, decide what area of your life you can borrow some time from. Maybe you could wake up a little earlier, or limit the time you spend on the Internet or watching television. Prioritizing math is not easy, especially at first. It may require you to make a drastic change in your lifestyle. If you genuinely want to do better in math though, you need to do this, and when you start to see the results of your efforts, you will be overjoyed.

Begin by creating a schedule that will allow you to do math everyday, preferably right after school. Then schedule math time into your planner, so you remember to do it (for example, write

"Do math 3:30-4:30" into your planner on every weekday). Then treat it with the same urgency as you would a job interview or a doctor's appointment—as if you absolutely cannot miss it. Every day that you spend the hour on math, put a checkmark or a happy face beside where you wrote it down in your planner. If you have four or five happy faces in your planner by the end of the week, for several weeks in a row, you will find yourself not only doing better in math but also feeling really proud of yourself.

Writing your "math appointment" in your planner is important because writing it down makes it real. Otherwise, improving in math will likely just stay in your mind as an idea or a "should". You may think you should spend more time on math, and you may even intend to work harder the following week, but next week often becomes next month and so on, and nothing changes. But if you have it written in your planner, and week after week you see no happy faces or checkmarks, it will start to bother you. On the other hand, you will likely be surprised at how happy you feel after completing the appointment even once or twice!

Of course, there is a challenge here: new habits are very hard to create. For at least the first couple of weeks, you will probably fight like crazy to get out of your math appointments, making excuses and finding reasons not to do the work. The good news is if you can get through that initial push, honoring your math appointment will get easier, until eventually, it will start to feel natural. One day when you get home from school, you will sit down and open your math book without even thinking about it. This is where we want you to be! The internal battle will be over. This is the beauty of persistence: eventually, the effort pays off and the habit sticks.

If, on the other hand, you cannot seem to develop a daily routine of working on your math, enlist help. This is a common struggle for students, and a good strategy is to ask your tutor, a friend, or a family member to spend your homework hour with you and keep you on task until you are able to do it on your own.

Another challenge for you may be that you want to spend the time working on math, but you seem to continually get stuck in the first few minutes. You may sometimes feel as though all the time in the world won't do any good. I know how frustrating this experience can be. The best solution is to start getting help with your math on a regular basis. Hire a tutor who you can meet with once or twice a week, or spend some extra time with your teacher.

3. **Pay attention to the lesson.**

Teachers often tell students to pay attention in class. Students hear it all the time, but I wonder if they actually know what it means. Paying attention is not by any means a passive activity; it does not mean sit there quietly and don't bother anyone. What it means is, listen and genuinely try to understand the material. Teachers want you to pay attention because *the lessons they give are for you.* Teachers create lessons so you can benefit and learn from them.

So listen to the lesson, get involved and try your best; a solid effort during class will result in less confusion and frustration later. If you do not understand something, raise your hand. If you are having a difficult time copying notes and listening at the same time, ask the teacher to slow down. If you seem to be having trouble understanding all the lessons, try reading the section for each day before class. If you look over the material before the teacher teaches it, and even understand it a little bit, you will probably find the lesson much easier to follow. Once you are able to follow the lesson and understand your notes, homework will become more manageable.

4. **Read your notes and textbook before starting homework.**

Suppose you were given a novel to read for English class, and the homework for the first day was to read Chapter 1 and answer the questions on a corresponding worksheet. How successful do you think you would be if you skipped directly to answering the questions without having read the chapter? Obviously not successful at all, because everything you need to know to answer those questions is in the reading. Math works the same way. You do get a lesson in math, but that is often not enough. Each day, before you start the homework, read over your notes from class, the corresponding section in the textbook, and all of the examples (pay close attention to anything in the textbook written in boldface or with a border around it). *Then* begin the homework; you will likely get through with much less frustration.

5. **Do more than just the assigned questions.**
    Remember that the purpose of homework is to give you practice so you learn the math. Your goal should not be to get the homework done; it should be to get it learned! So once you have completed the homework, if you are still a little unsure, do more questions. After all, practice makes perfect (or at least makes really good). I will go into more detail about how to do homework in the next chapter.

    A note about getting help: Often students will go to a teacher for help and the teacher will explain the concept in a way that the student does not understand. The student will tell the teacher that, and the teacher will explain it again the same way, except maybe slower. If this happens to you, ask your teacher to *explain it a different way*. Show him or her exactly where you are having trouble. The more specific you are about what you do not understand, the more able the teacher will be to help you. Try not to give up. Work with the teacher until you understand.

6. **Read the instructions!**
    "When all else fails, read the instructions!" Have you heard that before? Teachers often say it as a joke, but only because of the many students who have problems in math simply because they have not read the instructions. In fact, it is one of the most common reasons students lose marks on tests and have trouble with homework.

    For example, suppose a word problem requires the dimensions of rectangular garden, and the student answers x = 4. This does not answer the question; only one dimension is given (and we don't know which one) and there are no units. Or the instructions will say to simplify, and the student will factor instead, and then wonder why her answer is different from the one in the back of the book. Probably the most common mistake I see students make is not rounding to the required number of decimal places. The student will answer to three decimal places, for example, when they were supposed to answer to the nearest tenth of a meter. In all of these cases, simply reading the instructions carefully beforehand would have been an easy way to prevent incorrect answers or lost test marks. Reading instructions makes your life easier because the instructions tell you what you are supposed to do.

7. **Anytime you get a test or quiz grade lower than 100%, do the corrections.**

Doing corrections means looking over your incorrect and incomplete answers, finding out where you went wrong, and then doing the questions correctly. The purpose is to learn from your mistakes so that you can avoid making the same ones next time. Doing corrections does not mean asking your friend for the correct answer and copying it down. I know this can be tempting because it is easier, but you won't learn anything.

The best way to do corrections is on a separate piece of paper, which you then staple to the original test or quiz. If you erase your work on the original copy, you will not be able to see and remember where you went wrong—and you will be more likely to make the same mistakes again.

So start now. Open your binder, find the most recent quiz or test you wrote, and do the corrections the right way. Try to learn the math so well that if you were to take the quiz or test again, you would get 100%. It will be worth the effort; imagine how great you will feel when you see the question again and you know exactly how to do it!

8. **Start your homework as soon as you get it.**

As mentioned before, if you can get into the habit of doing your math homework right after school, your life will be much easier. If possible, do not even go home until you have finished it (or have done as much as you can without help). There are too many distractions at home, including the urge to unwind and relax.

Try this: when school is over, go to the school library or even the public library to do homework. If possible, find a friend in your class to go with you. I know doing homework can be hard right after school, but just think: if you can get your homework done first, then you can relax afterwards and do what you want for the rest of the evening. And you will have peace of mind from knowing you did the right thing. Wouldn't that be better than procrastinating all afternoon and evening, and then forcing yourself to start your homework at 9:00 at night, when you are exhausted from the day?

Note: Do this on Fridays too. I know on Fridays you probably think, "I have all weekend to do this; I don't have to worry about it now." But before you know it, Sunday night will have arrived, and your math textbook will be staring at you from your open backpack on your bedroom floor. If you can finish your homework instead by 5:30 or 6:00 Friday evening, you can relax and enjoy the rest of the weekend.

9. **Get organized.**
First, designate one binder just for math and organize it into three sections: one for notes, one for homework, and one for tests and quizzes. Keep everything, hole-punch everything, and then put everything in the right section of your binder. Also, label every piece of paper in your binder with a date and a title. Otherwise, you will not know what unit anything belongs to or when you did it. And remember, this is not because your teacher told you to, it is because *it will make your life easier.*

Second, get a planner if you do not have one already, and use it all the time—for *everything.* Write down every test date, dentist appointment, homework assignment, tutoring session, outing with friends—everything. Once you have everything written down, you can stop worrying and wondering if you have forgotten something. And when you open your planner to the current week, you will see exactly what you have to do and when.

Third, put systems in place to help you stay organized. For example, you could begin putting your keys in the same place every day when you get home. This way, you always know where they are. You could also organize your bedroom so that you have a designated place for all of your belongings. Again, this allows you to find everything easily.

Finally is the issue of supplies: bring all of your supplies to every single math class even if you are not planning to use them all: textbook, binder *with paper*, your homework, a pencil with lead, an eraser and a calculator. This will get you in the habit of bringing everything all the time, so that you always have what you need, when you need it.

Part of being in high school is learning to be responsible, and being responsible means having what you need all the time, especially on test days. If you show up to class on the day of a test or quiz without a calculator or a pencil, chances are your teacher will not lend you one, and she will probably get upset. And you don't want to waste other people's time trying to borrow whatever you forgot, or miss part of a lesson because you need to go to your locker. So develop the habit now of bringing everything, every time, no matter what.

10. Find a really great math tutor that you get along with.

If you are struggling in math, asking your parents to hire a tutor is one of the best things you can do for yourself. You will have encouragement, someone to talk to, and of course, help. You can ask the vice principal or a math teacher at your school to recommend someone, or you can look on the Internet or in the newspaper. Be selective though. If you find that the tutor is hard to understand or is not helping you, find someone else. Working with the right tutor will make math infinitely easier and more enjoyable.

If you cannot seem to find a good tutor, or if hiring one is not an option for financial reasons, find a way to get regular help at school.

11. Don't get behind.

If you miss a class or are unable to do your homework one night, catch up on the work you missed right away. Otherwise when test time comes, you will still have to learn the material that you missed *and* prepare for the test at the same time. This can be quite time-consuming and stressful, especially if you have trouble understanding some of the missed material.

If you miss class, get the notes and do the homework as soon as possible. Or if you are unable to do your homework one night because you are sick or have another commitment, make sure to do it the next day. You will be very grateful when test time comes around!

12. **Do not succumb to peer pressure.**

Several of my students have trouble concentrating in class because they have a friend in the class who constantly talks to them. I have other students who seem to have pressure all the time from friends wanting them to go out. The desire to fit in and have friends is healthy, but it can also set you back significantly if you are around the wrong people and allow them to influence you in the wrong way.

If you really want to succeed in math, make it clear to your friends that this is something you need to do for yourself. If they are truly your friends, they will understand and try to help. For you, this may mean overcoming the pressure to talk during lessons, to skip class, or to go out on the weekend when you have a test coming up. The temptation to go have fun with friends and not do your work can be overwhelming sometimes, and turning down a friend can be very difficult, so in these times, remind yourself of how badly you feel when you are doing poorly in math. Recall how much you don't want to go back to bad grades and low self-esteem. And remember that while any friend who may get upset with you will likely not stick around long, the self-esteem you gain from doing well in math will stay with you the rest of your life.

Chapter 8: More about Homework

I mentioned in the previous chapter that if you are struggling with math, the assigned homework is only a small part of what you should be doing each day. The students I tutor who do the assigned homework and no extra practice or review tend to forget the material very quickly. Then, when it comes time for them to review for the test, I am often reteaching previous sections. Sometimes it is as though these students have never even seen the material.

The truth is, the typical homework model in math does not work for most students; it allows them to finish their homework as quickly as possible and then forget about it. Unfortunately, this model often leaves students either not understanding the material, not retaining it, or both.

For these reasons, I am going to introduce you to a brand new approach to doing homework, which will keep you working on math everyday, whether you have an assignment or not. While that may sound difficult, keep in mind that it will be much easier once you have created the habit. You will be so happy when it comes time to study for your next test because you will already know most of the material!

A word of caution: keep in mind as you establish your daily study routine that spending more time on homework does not necessarily mean you are learning. What counts is *what you do* during that time.

How you spend your daily homework time will
greatly affect how you do in math.

So make sure that your extra time is time well-spent by following this four-step homework strategy, which will help you gain a better understanding of the material, improve on tests, and ultimately feel better about yourself.

Your Four-Step Homework Strategy
1. Read and learn the notes
2. Complete the homework
3. Check your understanding
4. Review

1. Read and learn the notes
Get into the habit of reading over your notes from class before starting the homework. Students usually forget parts of the lesson, or else they do not have time during class to absorb it fully. Going through the notes at your own pace can help ensure you understand the concepts and examples before leaping into the homework—and that should make the homework easier. Reading the relevant section in the textbook at this point also helps.

2. Complete the homework
Try to do the questions out of order to keep you thinking and engaged in what you are doing—you will learn so much more this way. If you finish the assigned homework and are still having trouble, or if you feel you need more practice, do a few more questions than you are required to.

3. Check your understanding
Look over your finished assignment and ask yourself, "If I saw these questions on a test, would I know how to answer them?" Then find out the answer by doing a practice quiz. Ask someone to write out a few of the questions from your homework for you in random order and then try them again. If you have trouble, you still need more practice. Always remember that the goal of doing homework is not just to get it done, but to learn the math.

4. **Review**

Once your homework is finished and you understand it, go back and review the previous sections in the unit. You could read over old notes, redo quizzes, and maybe even do a few randomly selected questions from each section. For example, after completing your 4.6 homework, do a few questions from each of the previous Chapter 4 sections, 4.1–4.5.

If a midterm or final exam is coming up in the next month or so, this is also a good time to begin reviewing for that, doing questions and reviewing notes from previous chapters. I have found that students who do not review regularly feel overwhelmed before a test because they basically have to relearn the material. If you spend a portion of every homework session reviewing, you will feel not only a lot less stressed, but also much more confident when it comes time to study for the test.

Note: If you do not have assigned homework on a particular day, go directly to Step 4.

As you begin to follow these four steps every day and watch yourself become amazing at math, keep in mind that sometimes things may happen that will prevent you from doing all of the steps every single day. Try not to worry too much about this; just pick back up on it the next day. The important thing is that you create the habit.

The main problems arise when students who struggle with math overload their schedules so that they cannot commit to this daily practice. I have seen students put themselves under more stress than most adults I know, playing soccer, taking singing lessons, doing chores, all while taking a full load of courses at school. Some students even get jobs. A schedule like this is unreasonably strenuous and difficult, and the student who tries to undertake it usually ends up either failing or dropping the math class, or feeling stressed and anxious about the class all year. So take care of yourself, and do only as much as you can handle.

Chapter 9: "But When Are We Ever Gonna' Use This Stuff?"

"Why are we learning this?"
"When are we ever gonna' use this stuff?"
"When will I ever need this in the real world?"

I have been hearing students ask these questions since I was in high school. And even after all this time, I have never heard an answer that seemed quite right. Usually teachers answer literally, giving specific examples such as, "Engineers use math all the time," "If you become an architect, you will need to understand trigonometry," or "You will need math to go to college."

The problem with those answers is that most students have not yet decided what they want to do after high school, and even if they had, probably a couple of students at most would have been planning to become an architect or engineer, while a handful more might have been sure to attend college. So, in other words, at the end of the day, most students still do not know how math is relevant for them.

Realistically, these students' questions are legitimate. The students are right: most high school math does not have a whole lot of real-world application. I mean, does an aspiring psychologist *really* need to know how to graph a cubic function?

A few years into my teaching career, I began to search for real answers to these questions, acceptable answers that would make sense to students and convince them that there really was a purpose to learning math, no matter what line of work they chose. And truth be told, I too wanted to know how high school math could be useful to *all* students in the adult world. So after giving the matter a lot of thought and speaking about it with several students and teachers, here is what I came up with.

To start, the math covered in early high school or middle school is extremely useful and practical. It involves skills that are applicable in the everyday world, such as working with fractions, percents, decimals, ratios, and integers. Students in these lower grades learn how to solve simple word problems including those that involve calculating discounts and tax. They also learn to estimate and to understand why estimating is useful.

So the question of how math is useful does not become truly tough to answer until the upper grade levels. Consider a Grade 11 lesson about quadratic functions, for example. When are you ever gonna' use *this* stuff? Well, the truth is you probably won't. Ever. Because let's face it: completing the square just isn't an important life skill. But you will use the skills and habits you develop during the learning process, especially if math has been difficult for you.

You see, working hard to succeed in math builds character. Through this experience, you will gain qualities that you can use way beyond high school, qualities that will make you stronger and smarter and more creative in life in general. Helping yourself in math forces you to become self-disciplined, responsible, organized, and an expert at time management. It also teaches you how to concentrate and stay focused.

After time, you will find that you are also better equipped to handle whatever life may send your way. You will learn how not to give up on something if you can't do it right away. You will learn how to get curious instead of frustrated when something doesn't make sense to you. You will learn that you can accomplish amazing things if you put your mind to it. And through all of this, you will mature, because when set your goals high and then work hard to achieve them, suddenly all of the little things you used to worry about seem much less important, and you will begin to think like an adult. You will gain an amazing sense of accomplishment that no one will ever be able to take away.

In addition, the math taught in higher grades reinforces skills and concepts learned in earlier years—concepts that are often applicable to everyday living. And your constant use of basic mathematical operations, for example, in addition to working with fractions, decimals, percentages, ratios, etc., will help you become *numerate;* that is, you will develop a sense of numbers and how they work (I will talk about this in depth in the next chapter). This basic numeracy is, in itself, worth working toward, because without it, you would likely feel anxious every time you had to deal with numbers—and in the real world, that is a lot.

For instance, I have been given the incorrect change in large chain coffee shops on several occasions, because the young adults working the cash registers do not know how to do basic math mentally. They just trust what the cash register tells them. Of course, this is probably fine most of the time, until something goes wrong—say, the employee enters the payment amount incorrectly (in the same way that you might enter the wrong number into a calculator), calculating the wrong amount of change due back to the customer. It is at this point that the young employees seem to go into a state of mild panic (and here they thought math anxiety ended after high school!). They usually say something like, "I suck at math" or "Sorry, I can't do numbers." Imagine how embarrassed these people must feel. The worst part about it is that people who do not become numerate will have to deal with this fear of numbers forever.

Finally, higher-level math enables you to exercise your problem-solving skills on a regular basis, as it involves thinking through problems carefully, estimating, and determining whether your answers are reasonable. And learning how to solve problems creatively is a trait that will help you in every aspect of your life.

So why exactly are you learning this stuff? The bottom line is that learning math is like lifting weights or working out. Just as people do these kinds of physical exercises to build strength and stamina, by really learning and doing math on a regular, consistent basis, you are building mental muscle and endurance. And the more often you correctly use and even put stress on these math muscles, the quicker, tougher, and more creative your mind—-and the stronger your spirit—will grow to be.

Chapter 10: The Gift of Numeracy

Simply put, *numeracy is the ability to comfortably and confidently work with, manipulate, and understand numbers in their different forms.* To help you understand what it means to be numerate, here is a list of some common characteristics and behaviors of numerate people:

Numerate people often will …

- do calculations in their heads using neat tricks (for example, to subtract 12 from a number, they subtract 10 first, then subtract 2 more);

- determine whether the answer to a problem is reasonable (in other words, whether an answer make sense);

- understand the meaning of large numbers (for example, knowing that 1 billion is 1,000 millions, which means that you would have to earn a million dollars a thousand times to be a billionaire!);

- count back change and know whether they have received correct change;

- estimate—and use estimation in their daily lives to help them figure things out;

- calculate approximate tax or tip amounts mentally;

- understand basic facts about numbers (for example, that $0.5 = \frac{1}{2}$ and that 1 and 1.0 mean the same thing);

- know basic units of measurement and roughly how to convert between them;

- understand the relationships between fractions, decimals, and percents, and when to use each;

- be able to calculate how much interest they will have to pay per year and per month on credit cards, mortgages, bank loans, car loans, etc.;

- have a general awareness of all the numbers in their world and how those numbers affect their life (for example, knowing how much of their money is going to interest or service charges, or knowing that a 10% discount on something does not even cover the tax); and

- trust their ability in math.

Numeracy versus Innumeracy

As you can see, there are many advantages to becoming numerate. And still, innumeracy seems to have become an epidemic among today's adolescents. I do not want you to become part of this new and frightening trend, because life is not easy for the innumerate; in fact, it can be quite frustrating at times, not to mention damaging to a person's self-esteem.

People who suffer from innumeracy struggle with basic tasks involving numbers, such as leaving the correct tip at a restaurant, determining how much tax they will have to pay on a new car, or figuring out that a line of credit with a 6% interest rate could save them thousands of dollars over a high-interest credit card.

Here is a perfect illustration of innumeracy: This morning, I went to a gas station to buy a coffee and a granola bar. The attendant told me the total was $2.05, so I gave him $10.10 (I didn't want $0.95 back in change, but a dime was the smallest amount of coinage I had). The clerk looked puzzled, gave me back $8, and then asked if that was the correct change. I said no, he owed me $8.05. I know the difference is only a nickel, but this sort of innumeracy happens all the time, and sometimes with much larger quantities of money, when it really does matter.

Developing numeracy can help you make intelligent decisions about important but basic money matters such as getting a mortgage, opening bank accounts, establishing credit, taking out loans, and even just shopping. You can also use it to interpret advertisements, loans, newspapers, statistics, sale prices, taxes, and the list goes on.

But the greatest part about becoming numerate is that your fear of numbers is replaced with curiosity about and fascination for them. You then gain mathematical confidence and independence, learning to figure things out without relying on anyone to help you. And the more you do this, the more you trust your own ability, until, gradually, you begin to develop a different way of seeing and interacting with the world around you. This is possible for you!

I love that I am numerate now. Now that I feel confident and comfortable with numbers, I am no longer afraid of math. Instead, I use math to figure out things, even just for fun. For example, the other day I watched a movie (For privacy purposes, I will not mention the title). In it, a man tells his psychiatrist that he is a genius and he just knows things, like that 197 times 546 is equal to 107,565. I knew immediately that this answer was incorrect. How, you ask? Because an odd number multiplied by an even number is always even! The correct answer to this multiplication is 107 562! (Note: These are not the exact numbers used in the movie; I just remember that one was odd and the other even and the "genius" came up with an odd number answer.)

This is interesting too: I recently saw a commercial that would fool most innumerate people. A pasta company came out with a new, whole-wheat version of their spaghetti noodles, and their commercial excitedly claimed that this new pasta contains 300% of the fiber of the original. Now consider what that might mean if their original pasta contained no fiber. It would mean the new pasta also contained no fiber, because 300% of 0 is 0. Or, say the original pasta contained 1 gram of fiber per serving—now you are getting a measly 3 grams! To me this is hilarious, but had I not been numerate, I would not have caught it at all.

Causes of Innumeracy

So if there are so many wonderful rewards to being numerate, then why do so many people still suffer from innumeracy? One reason is that innumeracy is socially acceptable. Listen to people talk, and you will often hear them joking about not being able to do math. Yet, how often do people joke about not being able to read? If someone cannot read or write or spell well, we label them as not smart, but if that same person does not understand basic math, we call them normal. What has happened in society to make being bad at math the norm?

Innumeracy might also be growing because so many innumerate adults are reluctant to change. That is because, regardless of the fact that innumeracy is socially acceptable, people who suffer from it often have low self-esteem due to their fear of math, especially if they had a negative experience in learning math. Many adults I meet, for instance, had such a traumatic history with their math education that they are now not only afraid but also adamantly against really learning it. They have associated math with so much pain that the easiest way to deal with it, they feel, is to avoid it altogether.

Another possible explanation for today's widespread innumeracy is that completing high school math does not guarantee that a student will become numerate; rather, numeracy is something a student has to work toward. As we have seen, numeracy comes from questioning and thinking and being curious. It comes from deliberately and habitually thinking about numbers and working with them in your mind rather than on your calculator. It comes from reasoning and estimating and second-guessing your answers. And it comes from doing all of this on a regular basis—even between math classes, and even after high school.

You Can Be Numerate!

Becoming numerate will help you to develop the confidence and skills and curiosity about numbers that will stick with you the rest of your life. The good news is that high school math is a four or five-year opportunity to train your mind to become numerate. If you approach math with curiosity, use your mind as much as possible instead of your calculator, practice estimating, and question whether your answers are reasonable, you too will become numerate!

~

It has been an absolute joy to write this book. I hope it has helped you feel more positive about math, and achieve the results you want. Most importantly, I hope it has showed you that you are smart and capable of success in math, and that you can achieve whatever you desire in this life if you work hard for it.

Now go study!

Afterword: A Note to Parents

Many parents I have talked with in my time as a teacher have expressed feelings of helplessness upon discovering that their child had been doing poorly in math. This seems especially true of parents who have their own bad memories of high school math. This chapter is devoted to all the parents out there who are feeling stuck. The chapter will present you with tools that you can use to first, understand how your child is doing in math, and then how to coach her so that she can improve and feel better. My hope is that these tools will empower you so that you can not only help your child, but also feel confident about helping her.

In my experience, students who are not doing well in math are usually in one of the situations outlined below. If you can identify where your child fits in, you will be better equipped to help her work through it.

> **Scenario 1: He doesn't seem to care.** To this student, math is not important. He may even be a bright student, but for some reason has no interest in focusing at school. He has other priorities.
>
> Students in this situation may exhibit the following behaviors: always claiming to have no homework, constantly on the phone or watching television, becoming introverted at home, not spending much time at home, and getting lower and lower grades. These students are usually not mature enough yet to see the value of working hard at school, or to think about the long-term consequences of their actions.

Students often lose their motivation to do well in math once they enter their teen years. They may suddenly have an entirely different set of priorities. At this stage of life, most adolescents enter into a highly social time. They want to have friends, be cool, and have fun. For some that might even mean getting involved with the wrong group of friends and starting to care more about fitting in than doing well in school—even if fitting in requires skipping class or using drugs and alcohol.

Scenario 2: She cares but is not trying hard enough. This child wants to do well in math but is not fully committing herself to it. This could be happening for any number of reasons. She may have too many other responsibilities or extracurricular activities, or maybe she really dislikes math and has trouble finding the motivation to do the work. It is also possible that she is simply unaware of exactly what is required to succeed in math.

Scenario 3: He is trying hard but still doing poorly. This student appears to be working hard—he may be in his room or at the library every night studying. And though he is probably genuinely trying, for some reason, he is still not doing well. It is likely that the student either does not understand the material, or is spending much of his study time unfocused. In either case, the majority of the time he spends studying ends up being unproductive.

Once you have identified what is going on with your child, ask her to tell you about it. Ask her what bothers her about math, and why *she* believes she is not doing better. Then be prepared for the answer; you may be surprised at what you discover. Often the issues run quite deep. Sometimes the student has never really understood math. Or sometimes her self-esteem is so low that the first priority should be doing small accomplishable tasks in math to build her confidence.

So learn about your child. Then the two of you can move on together to find the best solution (I will offer suggestions regarding how to do this later in the chapter).

The Deeper Problem: Trying to Find Creativity and Success in a Threatening Environment

Any of the above scenarios may have begun as a reaction to feelings of hurt, inadequacy, hopelessness, frustration, or anger toward math. These feelings often arise when a student is exposed to a negative or threatening environment for a prolonged period, usually either in a classroom led by an insensitive teacher, or at home where parents can get frustrated and act condescendingly in anger.

Such environments have caused an unacceptable number of students to believe that they are incapable of learning math, and I see evidence of this every day. Some teachers have emotionally scarred their students by not believing in them, criticizing them, and even going to such extremes as calling the student stupid. I am not sure these teachers are aware of how much damage this kind of behavior can do to a student's spirit.

In her book *The Thirst for Wholeness*, Christina Grof writes the following about what she calls intellectual abuse:

> During intellectual abuse the thinking process is disregarded, disrupted, or discouraged. For example, when people's ideas or thoughts are subjected to destructive criticism, when they are harshly judged or punished for errors in reasoning, when they are authoritatively and rigidly told how and what to think without room for creativity or error, they are being intellectually abused. (51)

The first time I read this description, I could not believe how well it seemed to fit the way adolescents are often treated by teachers and parents alike. I read part of the quote to one of my students and asked her what she thought it was describing, and do you know how she responded? She said *a bad math teacher!* I would like to clarify though: most of the math teachers I have met or worked with are caring, and amazing at what they do. The problem is that it only takes one or two insensitive teachers to seriously damage a student's confidence, especially if that student already doubts herself and her ability to learn math.

This form of abuse wreaks havoc on the creative process of an individual. Think about it. How could *anyone* have the courage to explore and learn and be creative in an environment in which he is constantly judged and criticized? It is no wonder that students who have been in such hurtful situations become unmotivated or unwilling to learn math. They end up afraid and unable to trust their own intuition. They become too scared to come forward with creative ideas. And they feel their self-worth is on the line, so they opt to stay quiet.

Then to make matters worse, parents often react to their children's low grades with anger. This can be devastating to a child who is already being put down by her teacher. Both school and home become places of hurt rather than of refuge, and the student is left alone, without support or help.

The result? Aside from the other personal problems this causes such as low self-esteem, these kids also grow to associate math with emotional pain, frustration, and even shame.

The Solution

If you do discover that there is a deeper problem, or even if you suspect it, your first step in helping your child in math should be to begin the healing process. A good way to do this is to read this book together. This will help the two of you connect and gain further insight into what the issues are. And you can use the book to reinforce the idea that your child's experience is common and that she is normal, which will, I hope, help her feel less alone. Then you can begin talking about success strategies such as forming good habits and developing persistence (see Chapter 3). The more you talk, the more positive and supported your child will likely feel.

Some General Guidelines

Here are some general strategies that should help you support your child in math. Following these strategies will require some of your time but if your child needs your support right now, the extra help and guidance from you may transform his attitude towards math. As a result, you may find him more eager to study and do homework, which will likely lead to a better report card.

The most important thing you can do (other than being supportive) is make sure you always have some idea of how your child is doing in math. If you wait until the first report card, it may be too late to correct the problem.

There are several ways that you can check his progress. First, encourage him to keep all of his notes, assignments, tests, and quizzes.

You can learn a lot about how he is doing by looking over these with him and going over any mistakes. Second, sit down with your child occasionally while he is doing homework, and watch him work. Ask him questions and try to get a sense of whether he understands the concepts. One great way to determine whether he is grasping the material, even if *you* struggle with math, is to ask him to teach you the lesson he is working on or show you how to do a question. If he does not understand the concepts, he will probably say something like, "I get it; I just can't explain it," or "I already understand it; I don't need to teach you." Usually students who understand the material are excited and will be eager to explain what they know.

Also, keep in mind that the assigned homework, especially for the struggling child, is only a guideline or a minimum of what the student needs to do to learn the material. The purpose of doing homework is for the student to learn the concepts and figure out how to apply them. To ensure that your child is in fact learning, every once in a while, after she has finished her homework, make a practice quiz for her to do. This can be as simple as copying some questions from the section she is studying in the textbook onto a piece of paper, and then checking her answers against those in the back of the book. If she does poorly, she has not learned the material well enough. In that case, have her go over the notes again and do more practice problems, then give her yet another quiz. You can continue this process until you are confident that she understands the material.

Of course, you will not always have time to sit down with your child while she is doing homework. On those extra busy days, at least take a minute to look over her homework and make sure that she has completed it. But be careful—teenagers can be sneaky! Some of them will just copy the answers from back of the book. Look for work, not answers.

Another quick way to check your child's progress is to call or e-mail the teacher regularly and ask for updates. Ask about your child's homework and grades and also about his behavior in class, as this is a great way to measure how he is feeling about math; students who are not doing well tend to act out or not pay attention.

These are all general guidelines for helping your child in math. But what if you encounter specific issues that deserve more tailored courses of action? What follows are some suggestions for more individual scenarios that may arise.

If the problem is due to lack of motivation
Helping your child get motivated may seem like an overwhelming challenge, but one technique that may help, if implemented correctly is introducing a *temporary* reward system.

Find out what your child values and then build the reward plan on that. For example, if she is trying to save money, or really wants to buy a new outfit, you could offer to give her a certain monetary amount if she scores 10% higher on her next test than she did on her previous one. This may really motivate her; after all, what teenager doesn't love money?

Once you have her attention, use the tools provided in this book to help her get started on the improvement path and then to coach her along. She will probably need a lot of support and guidance, especially if it has been a long time since she has tried in math. If she seems truly exasperated, this may also be the time to hire a tutor. In fact, you might even consider having a few tutoring sessions per week, just until your child gets caught up and begins to feel more confident.

If your child receives a higher grade than usual on the next test, she may be thrilled with herself despite the reward! But keep your promise and reward her—and then offer another reward for an even higher score next time. Continue using this reward system until she is doing *consistently* better, and then, at this point, you may want to have a discussion about the value of doing well for the internal rewards of confidence and pride. She may be motivated enough at this point to continue to try without external rewards. Imagine your child feeling so good about herself that doing well is enough to keep her moving forward!

If the problem is emotional
In an instance like this, the most important piece of advice I can give you is to be caring and encouraging. The "coach" and the child *both* need to work hard at maintaining a positive attitude about math.

Remember, math is a creative process, and people are rarely able to be creative when afraid or in an extremely negative state of mind. Your child needs to develop enough confidence that he has the courage to explore and experiment in math. He needs to feel *safe*. He needs to know that no one is going to laugh at him or criticize him or call him stupid if he makes a mistake. And it is especially important to maintain a safe environment at home if your child is experiencing intellectual abuse at school or even just having trouble with the teacher. Otherwise he may give up.

Another idea you may consider is taking your child to see a counselor or psychologist. Perhaps he is going through some emotional issues that pertain to areas outside of math and are preventing him from focusing on school in general. Maybe he is in the wrong peer group, for example, or is struggling with family issues. A counselor can work with him or the two of you on a regular basis, until things begin to turn around. If you decide to take this route, I still strongly recommend hiring a tutor; even when your child starts to feel better about things, his grades will not improve unless he is studying and doing homework effectively.

If the problem relates to lack of time

If your child has a full schedule, she may find it really difficult to come home after a long day and do math homework. Some of the students I tutor have volleyball, choir, or hours of dance practice every week, plus seven other courses besides math to worry about. In this case it may be necessary to make some sacrifices. Sit down with your child and make a list of priorities. Decide what is most important to both of you and then modify her schedule to accommodate the priorities. Be careful not to get fooled though; sometimes students have full schedules and still find an incredible amount of time to watch television, surf the Internet, or talk on the phone. If this is the case, the real issue is motivational, and a temporary reward system may be a better approach.

If the problem is not understanding math

If your child is consistently having trouble understanding the concepts, hire a knowledgeable and supportive tutor (see the Appendix to learn more about this). If you do not have the budget for a tutor, call the school and ask someone in the math department for suggestions. Some high schools offer free tutoring. Finally, follow the guidelines in this book so that you can help your child along the way.

Appendix: Helpful Hints for Parents

Finding the Right Tutor

If your child is struggling in math or lacking motivation, one of the first steps you can take, which gives almost immediate results, is hiring a tutor. The challenge is finding a tutor who is right for your child.

A good first step is phoning your child's school and asking if they can recommend someone. It is always safer to hire a tutor if he or she has been recommended to you.

Once you are in touch with a tutor, ask questions. Ask the tutor how many students she has tutored, and whether those students have improved. Ask her if she is a certified teacher, where she went to school, and what her specialty is. Ask her how familiar she is with the curriculum. You can even ask for references. Tutors are expensive, and your child's education is important, so you have every right to be fussy.

Be careful though because the tutor may sound amazing on the phone, and then end up not helping your child at all. Often this is because some tutors have a difficult time communicating or connecting with the student.

If you hire a tutor and find that it is not working out, be honest. Tell the tutor how you are feeling and what issues you see, and then either give her a chance to correct the situation or find someone else. Many families hold onto tutors out of fear that they will not be able to find a better one, or that they will hurt the tutor's feelings. But you can be honest without being hurtful—and who knows, maybe the tutor will learn something from the experience. People cannot improve unless they know what they are doing wrong.

The primary quality to look for in a tutor is a genuine caring attitude for the child and the child's success in math.

The ideal tutor will spend a good portion of the session listening to and guiding the student in the learning process. She will help the student come up with ideas and develop the courage to try new things. And she will teach the student not only techniques and problem-solving strategies, but also how to think about math.

In essence, look for a tutor whose overall goal is to teach the student how to learn math on her own—someone who tries to enable the student to get along without the tutor.

Several of my students have told me alarming stories about previous tutors they have had. One such student, whom I will call Kate, (I have changed her name here for privacy purposes) told me about a tutor who would actually do the homework for her. The tutor would write out the entire solution and then quickly explain it to Kate. Kate had no idea what was going on in the problem, and she felt as though the tutor was not even interested in whether she understood! And as if that were not enough, the tutor had small children in her home and was constantly leaving the room to tend to them.

Even after months with this tutor, the only thing that improved for Kate was her homework grades, and that was only because the tutor had already written out the solutions for her. Kate was still failing tests because the tutor was not teaching her the math or the skills to think about math; she was just doing the work for her. The experience left Kate feeling hopeless, as if she were destined to forever fail at the subject.

This is what can happen if you are not careful about whom you hire.

The Power of Collaboration

If you read about my experiences at the beginning of this book, you already know that Grade 12 was special for me because it was the first time in high school that I actually started to understand math. I believe this happened for two reasons: first, I had an excellent teacher for the first time, and second—and this is what I want to talk about here—I became "study buddies" with a classmate who shared my ability level. Having someone to work and struggle with really lightened the mood, and made studying less intimidating and frustrating. As we talked and laughed together, I even started to enjoy math a little!

Through this experience, and through my many observations as a teacher, I learned that most adolescents thrive in social environments. Once I realized this, I began thinking more and more about it until one day, I had an idea. I wondered whether the students I tutored would have a more positive experience with math if I grouped three or four of them together to do homework.

So I did an experiment. I created an additional program for my tutoring students called Homework Club. It has been running weekly for more than a year now, and the results have been amazing. The students are very productive and even happy, working for two solid hours without a break (other than to laugh a little or help one another). If these students were studying at home, I wonder if they would get half as much work done—or enjoy themselves half as much!

It was important to me to create a positive math atmosphere for my students, regardless of how many students I was working with at a particular time, because most of the students I tutored didn't even know such a thing could exist. They needed a place where they could feel safe to ask or be wrong, a place of "fearless math." And they needed a place where they could all help and support one another.

I am grateful to have these Homework Clubs because the extra time with my students allows me to focus not only on helping them with their math, but also on helping them change their attitudes about math and themselves. I have found that gradually, as we work together, these students break down the negative beliefs and feelings that they came to me with. The best part is that if they find themselves stuck on a question or concept, they have help on-site so they never have to get frustrated or feel alone. I spent hours feeling like that in high school.

You may want to encourage your child to try organizing a homework club, especially if she is in senior math. You can even hire a tutor to supervise and share the cost with other parents. If your child is serious about doing well, this may be a transformative experience for her.

References

1. Tobias, Sheila. *Overcoming Math Anxiety* (revised and expanded). W.W. Norton & Company, New York, 1993

2. Grof, Christina. *The Thirst for Wholeness-Attachment, Addiction, and the Spiritual Path.* HarperCollins Publishers, New York, 1993

3. Polya, George. *How To Solve It—A New Aspect of Mathematical Method* (second edition). Princeton University Press, Princeton, New Jersey, 1988

4. Wente, Margaret. "Go Figure, Ashley Can't." *The Globe and Mail.* May 5, 2001

Questions or comments for the author? Send an e-mail to visionformath@gmail.com